London

isk

J

B

C P

CIRIA *sharing knowledge ■ building best practice*

6 Storey's Gate, Westminster, London SW1P 3AU
TELEPHONE 020 7222 8891 FAX 020 7222 1708
EMAIL enquiries@ciria.org.uk
WEBSITE www.ciria.org.uk

Land contamination – management of financial risk

Finnamore, J R, Denton, B and Nathanail, C P

Construction Industry Research and Information Association

CIRIA Publication C545 © CIRIA 2000 ISBN 0 86017 545 6

Keywords

Financial risk, liability, contaminated land, risk management

Reader interest	**Classification**	
Landowners, trustees, landlords, tenants, developers, contractors, remediation technology vendors, financiers, investors, environmental and engineering consultants, chartered surveyors, loss adjusters, accountants, lawyers, underwriters, local authorities and regulators	AVAILABILITY	Unrestricted
	CONTENT	Guidance document
	STATUS	Committee-guided
	USER	All those involved directly or indirectly in land and property ownership, occupancy, transactions, development and remediation

Executive summary

Property represents an important and potentially valuable asset that can greatly influence a company's net worth. Commercial, industrial and other non-residential property assets in the UK were estimated to be worth approximately £515 billion at the end of 1998 (ONS, 1999). Businesses often rely on the value of their property to raise finance and attract investment; lending to the commercial property sector in 1999 was estimated at £65 billion, much of which will have been secured on the basis of the value of property assets (DTZ, 2000). Reduction in property value can thus impact on a company's viability and liquidity.

New and historic land contamination can seriously affect the value of property, even to the extent of turning an apparently valuable asset into an expensive liability. The risk of impairment depends on, amongst other factors, economic conditions (ie supply and demand) and the actual and perceived financial consequences of contamination (eg legal liabilities, remediation costs and restrictions in use of the asset).

Forecasting the risk has proved to be difficult due to considerable technical, scientific, legal and economic uncertainty. There has also been a lack of experience, or at least reported experience, in the UK of material losses on which to base financial risk analysis – although this position is gradually changing.

Uncertainty in the financial risk (rather than the risk itself) is the cause of much of the anxiety surrounding contamination in the UK. This concern is likely to have been compounded, on the one hand, by alarmist speculation. On the other hand, the buoyant property market (which has softened the impact of contamination in certain regions) and delays in implementing key pollution legislation have lulled some organisations into a false sense of security. What is clear is that the financial risk presented by contamination is real, as illustrated by several multi-million pound cases, and that whilst uncertainty remains, businesses should plan carefully their financial management strategies in accordance with the principles of sound corporate governance.

The traditional approach to financial risk management has involved a combination of technical solutions (eg due diligence, remediation and environmental management), contractual provisions (eg warranties and indemnities) and traditional insurance (eg public liability and property policies) which reduce risk and uncertainty, apportion liability, and finance losses respectively.

The effectiveness of these and other approaches has been brought into question by legal developments and changes in insurance, accounting and general business practices in the UK. Despite this doubt, companies have generally been reluctant to take additional steps to finance the residual risks before they materialise. Thus, in the event of a loss, the only option may be to make a charge against the balance sheet (profit and loss account). With the costs of contamination often running into tens of millions of pounds, businesses that are financially unprepared for these losses will be left in a vulnerable position.

Financial solutions are available to protect companies' business interests. Approaches include specialist environmental insurance, alternative risk transfer (ART) and self insured retention (SIR), many of which are used routinely to manage non-contamination risks. Their application to contamination risk is less well developed however, with the exception of certain industry sectors which have made specific provisions, built up contingency funds or established captive insurers. These and other financing techniques can effectively manage the residual risk that the more traditional technical, legal and financial methods of risk management fail to eliminate.

Acknowledgements

CIRIA's research programme on contaminated land aims to provide guidance for the construction industry, its clients and other interested parties and consists of a series of research and information projects dealing with the various aspects of treating and reusing land that is either derelict or contaminated or both.

Mr J R Finnamore of LGC (now with WSP Environmental), Mr B Denton of Abros and Dr C P Nathanail of Land Quality Management at the University of Nottingham wrote this report under contract to CIRIA.

Following CIRIA's usual practice, the research project was guided by a Steering Group, which comprised:

Ms V Fogleman (chair)	Barlow Lyde & Gilbert
Mr A Apps	Bovis Homes Ltd
Dr M L Athorn	BURA
Dr R Bailey	Loss Prevention Council
Mr D Baker	The House Builders Federation
Dr A Barker	FBE Management Ltd
Dr P Beckwith	British Waterways
Mr S Cook	AIG Consultants Ltd
Mr P Crowcroft	Environment Agency
Mr R Davies	AIG Europe (UK) Ltd
Mr K Deady	ASDA Stores Ltd, part of the Wal*mart Group
Professor T Donnelly	University of Newcastle
Mr D Graham	Royal Bank of Scotland
Mr G M Gray	CIRIA
Dr C Grundy	Arthur Andersen (formerly Mott MacDonald Group Ltd)
Dr M Keegan	National Power
Mr M Nash	Simmons and Simmons
Mr J Navaratnam	English Partnerships
Dr P Pritchard	Royal & SunAlliance
Mr D Rudland	Halcrow Group Ltd
Dr M Steward	Marsh Risk Finance
Mr M Taylor	Derbyshire County Council
Mr J Tuer	CA Blackwell (Contracts) Ltd
Mr P Wilbourn	Phillip Wilbourn & Associates
Mr G Withers	ECS Underwriting
Mr S Wood	BG plc

CIRIA's research manager for this project was Ms J C T Kwan.

The project was financially supported by AIG, Loss Prevention Council, CIRIA Core member organisations, BG Property plc, British Waterways, CA Blackwell (Contracts) Ltd, Construction Directorate of DETR under the Partners in Innovation programme, English Partnerships, Royal & SunAlliance, Royal Bank of Scotland, Shell Global Solutions, The House Builders Federation and Zurich Insurance Building Guarantee.

CIRIA and the authors gratefully acknowledge the support of these funding organisations and the advice and help given not only by the members of the Steering Group but also by Mr S Francis and Ms A Sanehi of DLA, Professor J Lowe of Parkman, Mr M Lowe of DETR, Mr P Surrey and the 92 individuals interviewed during the research.

User guide

This guidance document is divided into three sections.

Section A – Background describes the theory of financial risk management applied to land contamination issues. The section outlines the sources and causes of financial risk, the nature of uncertainty, the type of losses and their impact on business performance, the principles of risk management and the methods and approaches to protect a company's financial interests. A short review of current market practices is also presented, alongside forecasts of future trends.

Section B – Guidance provides a common approach to managing the financial risks of land contamination. The framework and methodology are described in Chapter 4 with users directed to Chapters 5–9 for specific information on the risks, liabilities and management options for the following activities:

Chapter	Activity	Intended users
5	Land ownership and occupancy	Landowners (including local authorities and homeowners) landlords, tenants, licensees, trustees, administrators
6	Transactions of assets or shares	Businesses directly involved in mergers, acquisition and disinvestment
7	Development and contracting	Developers, contractors, remediation technology vendors
8	Finance and investment	Financiers and investors
9	Provision of, and reliance on, professional advice	Clients, environmental and engineering consultants, chartered surveyors, loss adjusters, accountants and lawyers, underwriters, local authorities and regulators

Section C – Risk financing provides a review of financial management techniques, comprising retained risk financing, composite financing structures and insurance.

This publication addresses a complex and dynamic topic that encompasses a broad range of technical, legal and financial disciplines. It is intended for general guidance only. If any matter is sought to be relied upon, reference should be made to the appropriate adviser.

Contents

List of tables

List of figures

List of boxes

Glossary

Alternative risk transfer	Various non-traditional forms of risk financing usually including insurance and insurance derivatives combined within the same risk financing product as contingent capital and finite risk instruments.
Amortise	Pay off a debt using a sinking fund.
Appropriate person	A person who is liable for remediating contaminated land under Part IIA of EPA 1990.
Asset	Property or financial claim which can, if necessary, be converted into cash. Assets may be physical, such as property, or intangible such as markets, reputation and credit lines. Some assets may not always be recognised on the balance sheet of an entity, depending on its accounting policies.
Bond	A financially-banked guarantee from an individual or organisation, which is capable of being converted into cash on the occurrence of certain pre-determined events.
Brownfield	Land that has been previously developed (Kass, 1998). Not all brownfield sites are contaminated, or located in urban areas.
Captive	Insurance or reinsurance company owned by one or more insureds or others wishing to insure or reinsure the risks of their companies and/or their affiliates. Single owner captives may insure some or all of the parent company's risks as part of an overall insurance programme. Group captives usually have several owners or an association sponsorship and are originally formed to insure the risks of shareholders.
Caveat emptor	Let the buyer beware.
Collateral	Assets held to secure an obligation.
Consequence	Positive or negative result of an event.
Constructive directors (de facto directors)	Individuals who act as directors despite not being formally appointed.
Constructive obligation	An obligation that derives from an entity's actions where: a) by an established pattern of past practice, published policies or a sufficiently specific current statement, the entity has indicated to other parties that it will accept certain responsibilities and b) as a result, the entity has created a valid expectation on the part of those other parties that it will discharge those responsibilities (ASB, 1998).

Contaminated land

"...[A]ny land which appears to the local authority in whose area it is situated to be in such a condition, by reason of substances in, on or under the land, that:

a) significant harm is being caused or there is a significant possibility of such harm being caused or

b) pollution of controlled waters is being, or is likely to be caused"(EPA 1990; Section 78A[2]).

Contamination

Land that contains substances that, when present in sufficient quantities or concentrations, are likely to cause harm, directly or indirectly, to man, the environment or on other occasions to other targets (NATO CCMS, 1985).

Contingent asset

A possible asset for which its existence will be confirmed only by the occurrence of one or more uncertain future events not wholly within the entity's control (ASB, 1998).

Contingent credit

A form of committed financing whereby a lender undertakes to provide capital to a borrower contingent upon the occurrence of certain events. The capital might be in the form of shares, but more typically is debt or hybrid debt (redeemable preference shares or mezzanine loan finance).

Contingent liability

a) possible obligation arising from past events whose existence will be confirmed only by the occurrence of one or more uncertain future events not wholly within the entity's control

b) present obligation that arises from past events but is not recognised because:

- it is not certain that a transfer of economic benefits will be required to settle the obligation or
- the amount of the obligation cannot be measured with sufficient reliability (ASB, 1998).

Contra proferentem

Principle by which any ambiguity in contract wording is construed against the party who drafted the wording.

Controlled waters

All fresh and saline natural waters up to the UK offshore territorial limit, including rivers, streams, lochs, estuaries, coastal waters and groundwater. The statutory definition of controlled waters is given in the Water Resources Act 1991, s104(1) and COPA 1974 s 30A(d).

Corporate governance

Structures and systems by which companies are directed and controlled. Compliance with corporate governance principles may not be a statutory requirement.

Credit risk

The risk that a company may fail to perform its financial obligations.

Debt finance

A loan which is repayable usually with interest.

Deductible

First part of loss borne by insurance policyholder.

Derelict land

Land so damaged by industrial or other development that it is incapable of beneficial use without treatment (Wickens et al, 1993).

Diversification

A technique for managing risk where risk is divided among multiple, uncorrelated exposures.

Environmental management system	The part of the overall management system that includes organisational structure, planning activities, responsibilities, practices, procedures, processes and resources for developing, implementing, achieving, reviewing and maintaining the environmental policy (BS EN ISO14001, 1996, clause 3.5).
Equity	Share capital in a company.
Escrow	An account held independently of the entities which are party to the funds held in the account, so that the funds can be administered independently and in accordance with pre-agreed procedures.
Exculpatory contract	Contracts whereby the risk exposure of the transferor of the risk is eliminated.
Experience account	A reserve account established in a finite risk agreement. The balance of the account reflects the cedant's loss experience.
Exposure	Objects or situations where financial or other loss may be faced.
Financial risk	Combination of the likelihood of a financial loss (or gain) occurring and the magnitude of the loss (or gain). Risk may have positive or negative outcomes and reflects a deviation from a company's expected costs, income or asset value.
Finite risk	A limited liability insurance product for long-term cover, where arrangements are written into the contract to enable premium to be rebated to the insured usually by reference to the balance on an experience account.
Freehold	Land or tenement which a person holds completely and without restriction.
Hazard (risk factor)	Activities or conditions that create or increase harm/ damage (or benefit in the case of speculative risk). Financial hazards include activities (ownership/ redevelopment/transactions) or conditions (presence of suspected or known contamination) that create/or increase the likelihood of loss or the loss amount.
Indemnity	A contractual promise by one party to make good the loss suffered by another party or to pay a defined sum to that other party (Phillips and Ferera, 1998).
Insurance	A process of transferring the risk of loss from one party (insured) to another (insurer).
Internal control system	All policies and procedures that, taken together, support a company's effective and efficient operation by enabling it to respond appropriately to significant business, operational, financial, compliance and other risks to achieving the company's objectives (Turnbull, 1999).

Investment	A risk-taking financial commitment for which, in the event that the venture, project or business is declared insolvent, repayment takes a lower priority than claims by lenders and trade creditors
Liability	Obligations of an entity to transfer economic benefits or to conduct actions as a result of past transactions or events. Liabilities may be statutory, civil or contractual and may involve a claim upon a company's assets by another.
Liquidity	The ability to raise needed cash.
Long tail liability	Liabilities which may arise many years after the cause.
Mezzanine finance	Financial instruments which bear a blend of characteristics of debt and equity finance, namely that their financial yield may partly be dependent upon the profitability of the company or project, and may partly be secured against the assets in the event of default.
Mitigation	Limitation of the undesirable effects of a particular event (ISO/TMB, 1999 draft).
Multiplier yield	A property market yield which is equal to the rental value of the property divided by its capital value. The yield is a function of the covenant strength and the anticipated rental and value growth.
Peril	Cause of financial loss.
Piecrust lease	Lease that is limited to buildings, expressly excluding subsurface.
Pollution	The introduction by man into the environment of substances in sufficient quantity or concentration as to cause harm to human health, harm to living resources and ecological systems, damage to structure or amenity, or interference with legitimate uses of the environment (CIRIA 1997).
Portfolio	The holdings of more than one stock, bond, cash equivalent or other asset by an individual or institution. A portfolio may be designed to achieve an investor's goals, such as obtaining maximum returns, or reducing risk through diversification.
Principal	The capital element of a loan or investment.
Privity of contract	A principle that only the parties to a contract may receive benefits and incur liabilities under that contract; revised under the Contracts (Rights of Third Parties) Act 1999 in certain instances.
Provision	A liability of uncertain timing or amount, to be settled by the transfer of economic benefits (ASB, 1998).
Pure risk	A risk where only adverse outcomes are possible (compare to speculative risk).
Rating	The evaluation of the credit risk attaching to securities by an established rating service such as Moody's Investors Services and Standard & Poor's Corporation.

Remediation (a) The doing of anything for the purpose of assessing the condition of: (i) the contaminated land in question; (ii) any controlled waters affected by that land; or (iii) any land adjoining or adjacent to that land

(b) the doing of any works, the carrying out of any operations or the taking of any steps in relation to any such land or waters for the purpose: (i) of preventing or minimising, or remedying or mitigating the effects of, any significant harm, or any pollution of controlled waters, by reason of which the contaminated land is such land; or (ii) of restoring the land or waters to their former state; or

(c) the making of subsequent inspections from time to time for the purpose of keeping under review the condition of the land or waters (EPA 1990).

Residual contamination Amount or concentration of contaminants remaining in specific media following remediation.

Residual risk The risk remaining after risk treatment (ISO/TMB, 1999 draft).

Risk See **financial risk.**

Risk avoidance Decision not to become involved in a risk situation (ISO/TMB, 1999 draft).

Risk control Activities that focus on avoiding and reducing risks (both in terms of the probability and consequences).

Risk financing Activities providing the means to reimburse losses and to fund programmes to reduce uncertainty and risk. Prospective risk financing provides funds in advance of a loss, compared to retrospective financing which spreads payments across time periods following a loss.

Risk management Overall application of policies, processes and practices dealing with risk. This may include identification, analysis, evaluation, treatment, monitoring, review and communication (ISO/TMB, 1999 draft).

Risk perception Value or concern with which stakeholders and others view a particular risk (ISO/TMB, 1999 draft).

Risk reduction Actions taken to lessen the chance and/or negative consequences of a particular risk (ISO/TMB, 1999 draft).

Risk retention Acceptance of the burden of loss or benefit of gain from the consequences in the event of a particular risk. Retained risks will include those that have not been identified. Risks treated through insurance, or transferred by other means, are not retained, although the liabilities may be retained.

Risk transfer Share with another party the benefit of gain or burden of loss from the consequences of a particular risk (ISO/TMB, 1999 draft). Controlled risk transfer relocates the risk, compared to financial risk transfer which only transfers the responsibility to pay for losses.

Risk treatment	Process of selection and implementation of risk control options to manage risk (ISO/TMB, 1999 draft).
Securitisation	The process of converting rental income into bonds or other interest-paying IOUs that can be sold in the capital market (Catalano and Frampton, 2000).
Security	A lender's charge over the asset being financed. Secured loans may require a fixed or floating charge, mortgage or debenture over the property asset, perhaps allied to secondary security such as personal guarantees from directors or shareholders of the property owner. Security may also refer to the rating of an insurer or reinsurer.
Self-insured retention	Funding losses through the profit and loss account, and retaining rather than transferring risk.
Shadow director	Any person with whose directions or instructions the directors are accustomed to act. Excludes any person "by reason only that the directors of the company act on advice given by him in his professional capacity" [s721(2) Companies Act 1985].
Source of risk	Entity or activity with a potential to create impact.
Speculative risk	Risk of beneficial or adverse outcomes from an event (compare to pure risk).
Stakeholder	Any individual, group or organisation who may affect, be affected by, or perceive itself to be affected by risk (ISO/TMB, 1999 draft).
Stigma	That part of any diminution in value attributable to the existence of land contamination, whether treated or not, which exceeds the costs attributable to (a) the remediation of the subject property, (b) the prevention of future contamination, (c) any known penalties or civil liabilities, (d) insurance and (e) future monitoring (Syms, 1995).
Uberimmae fides	Utmost good faith – a duty imposed on both parties to an insurance contract to disclose all facts material to the contract.
Uncertainty	Situation where the future is not known and cannot be measured.
Warranty	An express or implied statement forming part of a contract that a particular state of affairs exists. It is normally qualified by a letter of disclosure which describes why, in certain cases, the warranty is not true.

Acronyms and abbreviations

ABI	Association of British Insurers
AGS	Association of Geotechnical and Geoenvironmental Specialists
ART	Alternative risk transfer
ASB	Accounting Standards Board
CAR	Contractors all risks insurance
CBI	Confederation of British Industry
CDM	Construction (Design) Management Regulations 1994
CGL	Comprehensive/commercial general liability insurance
CLAIRE	Contaminated Land Applications in the Real Environment
CPL	Contractors pollution liability insurance
D&O	Directors and officers
DETR	Department of the Environment, Transport and the Regions
EA 95	Environment Act 1995
EC	European Community
EIL	Environmental impairment liability insurance
EL	Employers liability insurance
EMAS	Eco Management and Audit System
EMS	Environmental management system
EPA 90	Environmental Protection Act 1990
EU	European Union
FOCIL	Forum on Contamination in Land
FRS	Financial reporting standard
FTSE	Financial Times Stock Exchange
HSE	Health and Safety Executive
ICAEW	Institute of Chartered Accountants in England and Wales
ICE	Institution of Civil Engineers
IPC	Integrated Pollution and Control
IRR	Internal rate of return
ISO	International Standards Organisation
LA	Local authority
LCR	Land condition record
LPA	Local planning authority
M&A	Merger and acquisition

PFI	Private Finance Initiative
PI	Professional indemnity insurance
PL	Public liability insurance
POST	Parliamentary Office of Science and Technology
PPC	Pollution Prevention and Control
PPG	Planning policy guidance
PV	Present value
RDA	Regional Development Agency
RICS	Royal Institution of Chartered Surveyors
SEPA	Scottish Environment Protection Agency
SIR	Self insured retention
SPL	Significant pollutant linkage
SSSI	Site of special scientific interest
T&M	Time and materials
TCPA	Town and Country Planning Act 1990
UKAS	United Kingdom Accreditation Scheme
UST	Underground storage tank
UTF	Urban Task Force
WRA 91	Water Resources Act 1991

SECTION A – BACKGROUND

1 Introduction

Potential land contamination liabilities in the UK stood at twenty billion pounds according to an estimate by the Confederation of British Industry in the early 1990s (Confederation of British Industry, 1993). More recent research from the former Department of the Environment estimates the figure to be nearer fifteen billion pounds (ENDS Report, 2000). Whilst considerable uncertainty surrounds the actual cost to business, it is clear from the sizeable provisions disclosed in the statements of several publicly quoted corporations that sections of industry recognise contamination as a major business issue (see Table 1.1).

Table 1.1 *Examples of corporate financial provisions for contamination liabilities, 1997 (Doolittle, 1999)*

Company	Provision (£m)	Comments
British Gas	342	Projected costs of "statutory decontamination"
Royal Dutch Shell	412	Covers environmental clean-up liabilities

Despite the potential scale of contamination liabilities and a heightened awareness of the risks, a number of organisations remain inadequately protected against financial loss. These organisations risk suffering significant distress in the event that a loss occurs. Other companies that invest too heavily in measures to control contamination liability risk suffering increased losses elsewhere in their operations. In both cases, business performance suffers.

1.1 OBJECTIVES AND SCOPE OF REPORT

The purpose of this report is to assist organisations to manage the financial risks posed by land contamination. It aims to do so by:

- highlighting the financial risks and liabilities to which businesses are exposed
- outlining the range of techniques and approaches that can be used to manage risk
- summarising the benefits and limitations of management techniques
- providing a structured framework and methodology to develop appropriate risk management strategies.

This report provides guidance on managing the financial risks of land contamination in its broadest context and not just contaminated land as defined under statute (see Section 1.2.1). It is targeted at the wide range of organisations – large or small, industrial or non-industrial – which are directly or indirectly exposed to these risks in the course of their business activities. The guidance advocates an approach to risk management that seeks to:

- reduce a company's risk of acquiring new contamination liabilities
- treat existing risks and liabilities at an appropriate cost to the business
- ensure that a company is prepared financially for residual risk
- integrate contamination risk management practices into a company's overall risk management programme.

Whilst the guidance highlights technical, legal and financial management options, the focus of the report is on financial techniques.

Consultation

This report is the culmination of three phases of consultation with organisations involved directly or indirectly in land ownership and development in the UK. In the first phase, a total of 455 questionnaires were sent to companies in the following sectors: accountancy, contracting (construction and remediation), consulting (engineering, environmental and financial), conveyancing, finance (banks, building societies, investors and fund managers), insurance and reinsurance (including underwriting agencies and brokers), laboratory services, land development and regeneration, legal, property holding (including freeholders, tenants and licensees), receivership and regulation (Environment Agency, Scottish Environment Protection Agency [SEPA] and local authorities). The survey provided an overview of financial risk exposure and management practices in each of these sectors. The response rate was 40 per cent (181 replies), with organisations from all sectors responding.

The second phase of consultation consisted of detailed interviews with 92 of the respondents. Activity-specific interview scripts were used throughout, enabling an in-depth assessment to be made of the specific financial problems faced by sectors of UK industry and the range of risk management solutions available to, and used by, businesses. Finally, a follow-up workshop was used to present and test the findings of the consultation process.

UK legal position

The legal regimes in the four nations that make up the United Kingdom have many areas in common, but they are by no means identical. The devolution of environmental issues to the Welsh Assembly has introduced disconformity with English Law, which may widen with time. For example, the implementation of Part IIA of the Environmental Protection Act 1990 took place earlier in England than Wales and Scotland. Since the EPA 1990 does not apply in Northern Ireland, statutory nuisance is dealt with under the provisions of the Public Health (Ireland) Act 1878. CIRIA's SP112 report (1997) provides an overview of the legal aspects of land contamination, including the differences between the four nations. However, that report was produced prior to devolution to the Welsh Assembly and Scottish Parliament.

1.2 BUSINESS CONTEXT

1.2.1 Contamination and business risks

Companies are exposed to a multitude of business risks of which environmental risk has become an increasingly important component. In turn, it is widely recognised that land contamination can contribute substantially to organisations' environmental financial risk exposure. Part IIA of the Environmental Protection Act 1990 (EPA 1990) addresses a subset of those contaminated sites where the degree of contamination presents unacceptable harm or risk of harm under the current use. Such sites are defined under statute as contaminated land (see Section 2.4.2). The UK Government intends that the planning regime will continue to drive the remediation of the remaining contaminated sites.

Figure 1.1 illustrates the relationship between contaminated land, land contamination more generally, and a company's environmental and business risks.

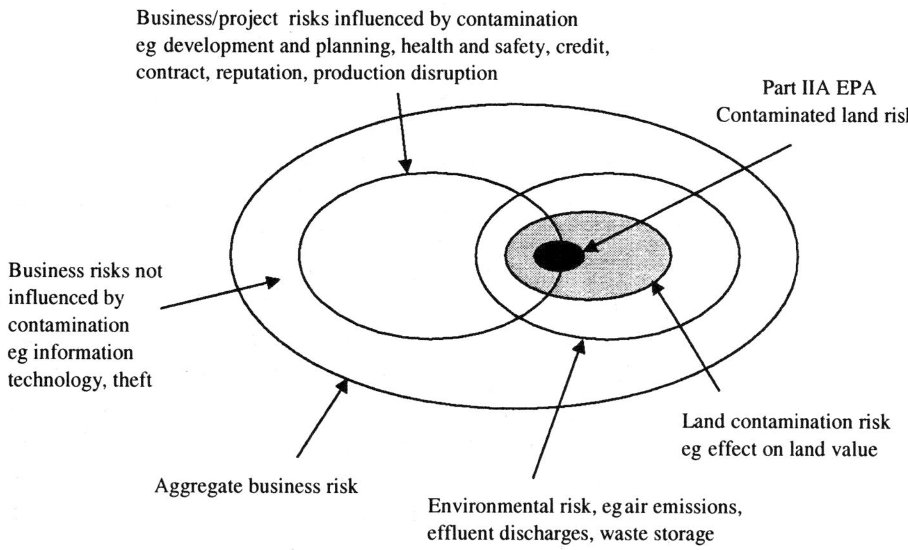

Figure 1.1 *Relationship between contamination, environmental and business risks*

In recent years, contamination as a business risk has commanded greater attention at a corporate level as a result of several technical, financial, legal and business developments, including:

- introduction in 1998 of a new accounting standard, Financial Reporting Standard 12 (FRS12) *Provisions, contingent liabilities and contingent assets* by the Accounting Standards Board (ASB, 1998). One impact of FRS12 is the requirement for companies to recognise certain liabilities for remediating contaminated land in their financial statements
- adoption (as the basis of sound corporate governance) of a Stock Exchange *Combined code* (London Stock Exchange, 1998) which has marked a further move towards protection of stakeholders' financial interests through defensible and transparent risk management
- from 3 July 2000, a requirement for pension funds to disclose their socially responsible investment (SRI) policies, including how social, environmental and ethical issues are taken into account in investment decisions (Nicholls, 2000)
- restrictions of pollution cover on virtually all public liability and some other insurance policies written after 1990 in the UK
- introduction of strict and retroactive liability for remediating contaminated land (Part IIA of EPA 1990)
- introduction of a Pollution Prevention and Control (PPC) regime which will bring with it financial implications associated with remediation on relinquishing a permit
- Government moves to reduce development pressures on greenfield sites and encourage greater development on brownfield sites, some of which may be contaminated
- a number of well-reported contamination incidents which caused significant financial impact on the organisations and individuals involved (see Box 1.1)
- several contamination-related disputes which resulted in multi-million pound civil actions
- a series of less well publicised, financial losses incurred by financiers and investors as a result of known or suspected contamination
- increasing evidence and public awareness of the damaging effects that contamination can have on humans, ecosystems, water resources, the built environment and businesses.

Box 1.1 *Examples of incidents and disputes involving land contamination*

Weston Quarries, Runcorn, Cheshire. Drilling at Weston Quarries, not far from ICI's Runcorn site, established there had been movement of hexachlorobutadiene (HCBD) – a waste product that had been disposed of at this location over a period of 50 years. The disposal had stopped approximately 25 years ago. Indoor monitoring of some 130 properties found 22 buildings with some presence of HCBD. Residents were evacuated and ICI agreed to underwrite the value of affected properties for a number of years (*Panorama*, BBC Television, 2000).

Lumsden Road, Portsmouth. In the early 1990s, approximately 120 families were re-housed after it was discovered that part of a housing estate leased to Portsmouth City Council had been built on a former Royal Navy waste site which was contaminated by asbestos and heavy metals. Re-housing, medical screening, demolition, remediation and rebuilding cost in the order of £6 million (Parliamentary Office of Science and Technology, 1993).

Ilford, Essex. In the late 1980s, low-level radioactivity was encountered on a site being redeveloped for housing. This caused significant project delays and required remediation at an additional cost of approximately £10 million (Sury and Slingsby, 1992).

Dungeness Power Station, Kent. A leak in the fuel supply to the Dungeness Power Station allowed an estimated 190 000 litres of oil to seep into the underlying shingle aquifer. British Energy were fined £70 000 for the groundwater pollution, spent £0.5 million investigating the incident and were expecting to spend another £1 million on remediation (*ENDS Report 289,* February 1999, page 53).

Loscoe, Derbyshire. In 1986 an explosion destroyed a bungalow, badly injuring the occupants. The cause of the explosion was methane migrating from a nearby landfill site. The after-effects included difficulties securing mortgages and insurance (Derbyshire County Council, 1986).

Bonttdu, Wales. In 1999 the proprietor of a service station was fined after an underground storage tank caused pollution of a river in Bontddu. In 1997 during installation of a remediation system, petrol vapour was disturbed in the ground, resulting in a build up of vapour in a nearby cottage, which in turn led to an explosion. The Health and Safety Executive (HSE) fined the consultants managing the works.

Wimborne, Dorset. In February 1999, Esso was fined after a leak from one of its service stations resulted in about 7000 litres of unleaded petrol polluting a major aquifer near Wimborne, Dorset. It was reported that approximately 1000 litres had been recovered, but the fate of the remainder was unknown. The clean-up had cost £120 000 to February 1999 while removal and replacement of the service station storage tanks cost £380 000 (*ENDS Report 289*, February 1999, page 54).

Urban Regeneration Agency v Mott MacDonald Ltd (1998)

£18.5 million damages were awarded for negligent advice provided by the developer's consultants. An appeal was made and the case was settled out of court before the appeal was heard.

Oil pollution from railway land. Railtrack and Silver Link train services were fined £125 000 each in February 2000 after admitting polluting groundwater with over 200 000 litres of oil. Clean-up costs were estimated to exceed £1 million (*ENDS Report 301*, February 2000, page 50).

Blue Circle Industries v Ministry of Defence. In July 1989, a storm caused ponds at the Atomic Weapons Establishment (AWE) at Aldermaston to overflow into a stream which drained through marshland on to land owned by Blue Circle. The marshland became contaminated with radioactive material, including plutonium, which it was not practically possible to separate from the soil. Radioactivity levels were well above normal background levels and in excess of statutory limits, but below levels which would have posed a risk to health. Expert valuers agreed that the economic value of the entire estate had been affected. Furthermore, it was considered that the estate could not be sold until the contaminated soil had been removed. Blue Circle brought a civil action under the Nuclear Installations Act 1965, seeking damages for the loss it had suffered. The company was awarded damages of over £6 million (Toulmin, 1998).

1.2.2 Business opportunities

Land contamination is seen as an unwanted burden by many organisations. Others have seen it as an opportunity to generate new business, typified by the insurance industry introducing a range of solutions to manage contamination risk. Development of contaminated sites has also become an important business niche for certain developers and a new speculative business venture for some consultants and engineers. These companies have recognised that remediation of contaminated sites can, in certain cases, catalyse regeneration of derelict land, introduce value and bring wider benefits to the local community. For example, it was reported that the decision of Derby County Football Club to relocate to the Pride Park redevelopment after remediation was instrumental in attracting other businesses to the site (Braithwaite, 1999).

Owners of land portfolios can improve business performance by targeting capital and resources to high priority sites yielding greatest returns, and by improving the value of their portfolio by selective site acquisition and disposal. An effective, ongoing risk management programme may also offer a competitive edge by providing a company with the tools to plan for, and respond effectively to, future UK and EC legislation.

1.2.3 Accounting for contamination

FRS12 provides the first detailed accounting standard for recognising, measuring and disclosing environmental liabilities in the UK. Before the standard was issued, there was more scope for "flexible" interpretation of the impact of contingent assets and liabilities on a company. This enabled businesses to reduce the impact of their contingent liability or reflect it in a period most suitable to the company.

FRS12 is designed to protect stakeholders' interests by requiring a transparent valuation of an organisation's true net worth, ie by reflecting the present value (PV) impact of contingent liabilities and assets (Gardner, 1999). The Standard requires that companies put a balance sheet provision or contingency to cover their contingent liabilities (including environmental liabilities) which in turn must be disclosed in the financial statements. Provisions will only be recognised where a present obligation – legal or constructive – arising from a past event will in all likelihood give rise to a transfer of economic benefit. This makes it more difficult for companies to smooth profits by assigning large, one-off provisions against which future expenditure can be charged.

In terms of land contamination, FRS12 will focus companies' attention on their present and future obligations, such as the costs of remediation and third party damages arising from past polluting activities. Obligations arising from decommissioning and claims under contamination warranties will come under similar scrutiny. Provisions that are created to cover the costs of contamination are reflected in present value terms. Therefore, where the costs are estimated as arising at a period well into the future, an organisation may not need to provide for a significant amount in the accounts. In addition, where an organisation continues to derive economic benefits from a potentially contaminating activity, the financial statements may recognise the entry to the provision for cleaning up the contamination as an asset. This asset may then be written off over the remaining period from which economic benefits are derived.

FRS12 provides little guidance on how to quantify land contamination liabilities and make appropriate provisions. Given the considerable technical and legal uncertainties surrounding land contamination risk (see Section 2.3), there remains scope for broad interpretation of the standard. It remains to be seen how accountants and directors will tackle the uncertainties and intangible losses that are associated with environmental risk.

1.2.4 Corporate governance

The combined code

Poorly managed business practices can create major corporate failures, as shown by the demise of the Bank of Credit and Commerce International and Polly Peck in the 1980s and the Maxwell Group companies in the 1990s. In an attempt to prevent the recurrence of such business failures, the UK adopted *The combined code, principles of good governance and code of best practice* (London Stock Exchange, 1998) based on internal controls designed to protect shareholders' investments and companies' assets.

The combined code (referred to here as "the Code") was developed from the Cadbury, Greenbury and Hampel reports on the financial and directors' remuneration aspects of corporate governance. It should be read in conjunction with the Turnbull report (Turnbull, 1999), which provides guidance on implementing the internal control requirements of the Code.

Whilst the Code is directed at corporations listed on the London Stock Exchange, the underlying principles reflect good business practice and are, therefore, more widely applicable. It has become recognised that improved standards of corporate governance, including environmental management, are necessary for organisations to remain competitive, attract external investment and provide a basis to comply with stricter regulations and legislation. Many companies have reacted accordingly by developing environmental policies and committing to the Code.

The Turnbull report

Internal control systems are recognised as playing a key role in the management of risks that are significant to the fulfilment of a company's business objectives. The Turnbull report indicates that a company's internal control system should be embedded within its operations and should include policies and procedures to assess risks, procedures to respond to risks and procedures for reporting identified weaknesses. Directors are expected to review regulatory reports on the effectiveness of the system of internal control, and to undertake an annual assessment for the purpose of making their statements on internal control in the annual report.

1.3 BROWNFIELD INITIATIVES

The Government has committed itself to establishing an "urban renaissance" which encourages urban regeneration, minimises unnecessary greenfield development and helps create more sustainable patterns of development (Urban Task Force, 1999). Among the measures that have been agreed is a sequential approach to housing planning for the 3.8 million new houses which projections suggest will be needed in the UK by 2016. The approach requires brownfield sites to be prioritised for development in advance of greenfield sites, and sets a national target that 60 per cent of new dwellings should be built on brownfield sites or through recycling of buildings.

The Urban Task Force (UTF) was assembled by the Government in 1998 to evaluate the factors that have caused urban decline in England, and to recommend practical solutions to turn urban areas into places where people want to live and work. Much of the UTF's work focused on identifying mechanisms to secure greater participation in the development of urban areas and the conversion of redundant buildings and recycling of contaminated sites to deliver "urban renaissance".

The UTF reported its findings in 1999 (Urban Task Force, 1999). It concluded that, based on the policies applying at that time, the target of 60 percent of new houses on brownfield sites would not be achieved unless there was a material change in planning and other Government policies, including:

- limiting greenfield releases and channelling development towards brownfield sites
- releasing redundant urban land and buildings for regeneration
- establishing a national framework for identifying, managing and communicating the risks that arise throughout the assessment, treatment and after-care of contaminated and previously contaminated sites
- setting up an Environment Agency "one-stop shop" service for regulatory and licensing requirements, moving quickly to a situation where a single regeneration licence is available and covering all regulatory requirements for cleaning up a site
- providing standardised Land Condition Records to offer more certainty and consistency in the management and sale of contaminated land
- establishing a national campaign to bring all land affected by contamination back into beneficial use by 2030.

Historically, public funding has been available to bring brownfield sites in selected areas back to a condition where they can be redeveloped through private finance. The Derelict Land Grant (DLG), the Garden Festivals and the work of the Development Corporations in the 1980s stimulated economic activity in many former industrial areas. This continued into the 1990s through the City Challenge Initiative and the work of the Welsh Development Agency, Scottish Enterprise and English Partnerships nationally, and the newly formed English Regional Development Agencies (RDAs). Even before the Development Corporations were established, the Government had piloted other regeneration initiatives, such as the Commission for New Towns, to secure development on previously used land.

The demand for brownfield development sites has increased against a backdrop of technical, cost and remediation difficulties. Contaminated Land Applications in the Real Environment (CLAIRE) was established in 1999, with the intention of stimulating the development of cost-effective methods of investigating and remediating contaminated sites in the UK (CLAIRE, 1999). CLAIRE is a public/private partnership involving stakeholders from industry, academia, regulators and the Government. A number of other initiatives are also operating in the UK and across Europe. These share similar aims of testing land remediation techniques and proving new approaches. The initiatives tend to be based on a network of independently owned test sites, which are made available for research and development in new technologies, together with their demonstration and verification. It is hoped that by providing test sites to demonstrate the cost-effectiveness of technologies under UK conditions, overall confidence in the technical and financial aspects of full-scale remediation will be increased.

Perceived inconsistencies by the UK market in the measurement and management of risk (and the financial consequences) point to the need for a new national standard for recording and communicating the risks of land contamination. The recommendation of the UTF is to introduce a standard format for recording site information such as historic land use, contamination and remediation activities. This could be set out as a Land Condition Record (LCR). Government could also introduce a "kitemark" certifying the quality of the management methods adopted by site owners, developers and their representative organisations. In recognition of the importance of assessing the potential risks of land contamination during property transactions, guidelines for the preparation of LCRs have been developed and their use in practice is being piloted.

2 Risks and liabilities

2.1 ELEMENTS OF FINANCIAL RISK

Businesses operate with an underlying aim to enhance their financial performance by maximising profit and return on capital. Financial risk encompasses the threat (or opportunity) of a company failing to meet (or exceeding) its business objectives. The risk is the combination of the likelihood of a financial loss (or gain) occurring and the magnitude of the loss (gain), and is seen as a variance from a company's expected costs, income and asset value.

Risk can have favourable and unfavourable effects depending on whether there is a downside or upside variation in cost, income and asset value. Pure risks have only unfavourable outcomes, whilst speculative risks generate beneficial or adverse outcomes. For the purposes of illustrating how an organisation may suffer loss as a result of known or suspected land contamination, the financial risk is broken down into constituent elements (as shown in Figure 2.1) and analysed in Sections 2.2–2.7.

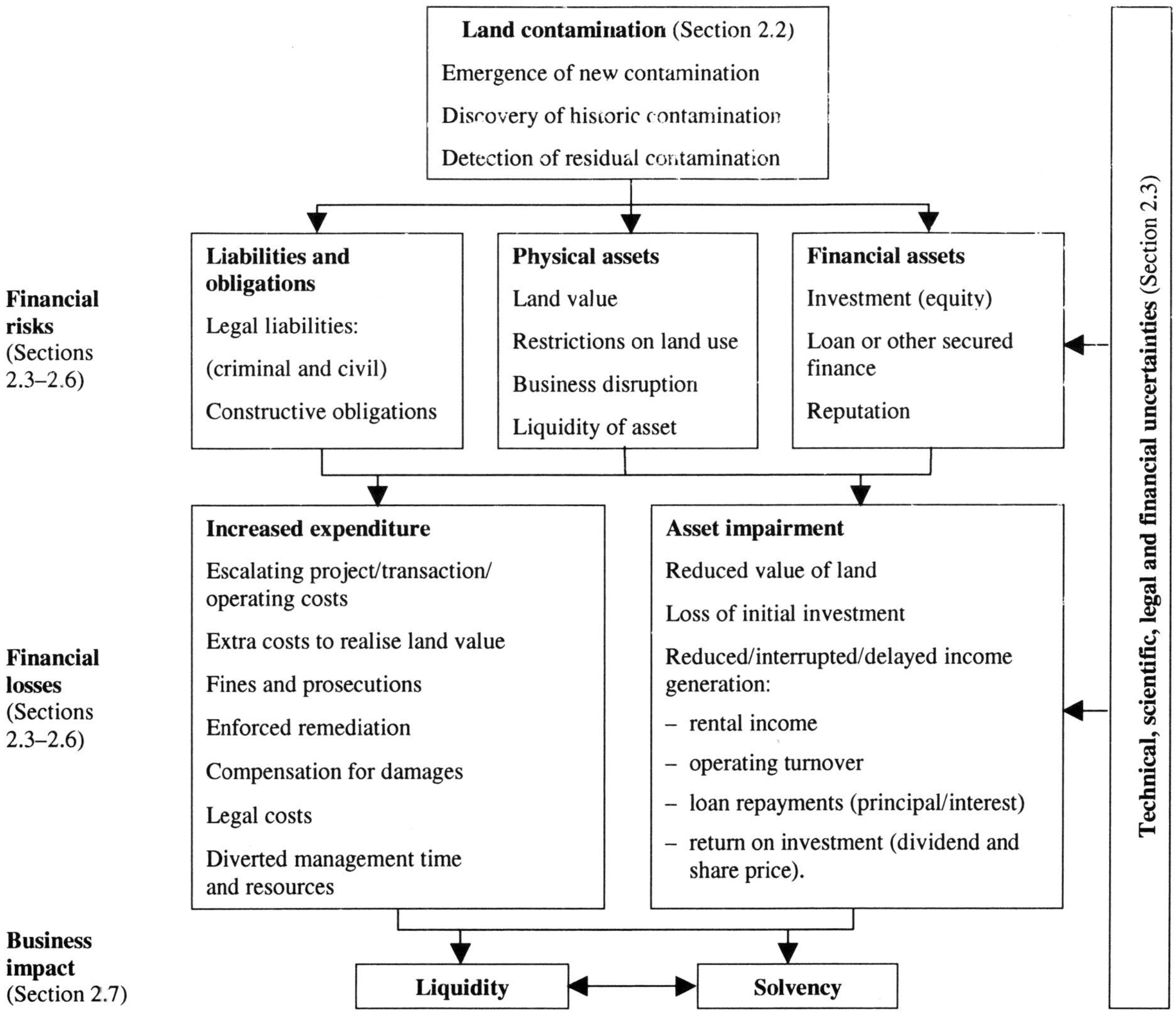

Figure 2.1 *Elements of contamination-related financial risk*

2.2 LAND CONTAMINATION

The financial consequences of land contamination can be traced back to the actual or perceived physical harm that it can cause. Harm may only occur where a receptor is exposed to a hazard, such as a contaminant. In turn, exposure requires a pathway to link a contaminant to a receptor ie a contaminant–pathway–receptor linkage (pollution linkage).

Box 2.1 *Definition of land contamination and associated terms*

Land contamination: land that contains substances that, when present in sufficient quantities or concentrations, are likely to cause harm, directly or indirectly, to man, the environment, or on other occasions to other targets (NATO CCMS, 1985).

Contamination may be chemical, biological (eg anthrax), radiological or physical (eg concrete foundations). Chemical contamination may be present in various states (eg gaseous, dissolved, liquid, solid), species (eg chromium (III) and (VI)), and forms (eg bioavailable).

Contaminant: a substance which is in, on or under the land and which has the potential to cause harm or to cause pollution of controlled waters (EPA 1990)

Pathway: one or more routes or means by, or through, which a receptor is being exposed to, or affected, by a contaminant, or could be so exposed or affected. (EPA 1990)

Receptor: entity or object having the potential to be harmed by a contaminant

Hazard: a property or situation with the potential to cause harm, eg the presence of a contaminant

Suitable for use: an approach to remediation that recognises that the risks presented by any given level of contamination will vary on a site-specific basis according to the use of the land and a wide range of other factors. The approach consists of three elements (EPA 1990):

a) ensuring that the land is suitable for its current and likely use not requiring planning permission

b) ensuring that land is made suitable for any new use, as official permission is given for that new use

c) limiting requirements for remediation to the work necessary to prevent unacceptable risks to human health or the environment in relation to the current use or officially-permitted future use of the land.

2.2.1 Sources and causes of land contamination

Past industrial activities, such as waste disposal and mining, have left the UK with a legacy of contamination. This can be attributed to point sources, such as spillages, waste deposits and effluent discharges, or more diffuse sources, such as the widespread application of pesticides and sewage sludge, or the aerial deposition of particulates. Contamination is also found naturally, eg radon gas (emitted from rocks from the natural decay of uranium and thorium), methane (from the decomposition of organic materials), organic compounds (generated following forest fires), natural oil seepage (from coal and oil shales) and toxic elements (occurring in rocks).

The number of contaminated sites in the UK has been estimated at 50 0000–100 000, equating to a land area of 100 000–200 000 hectares or 0.4–0.8 per cent of the total UK land area (POST, 1993). More recently it has been suggested that there may be 300 000 hectares of land in the UK affected to some extent by industrial or natural contamination (DETR, 2000).

Existing activities, licensed or not, have the potential to add to or exacerbate historic contamination. This can occur, for example, as a result of plant malfunctions, such as leaking underground storage tanks. Activities at a site may also mobilise existing contamination or introduce new pathways or receptors, thereby creating a new pollutant linkage. Even sites that have undergone remediation are likely to have some residual contamination.

Contamination beneath a site may not necessarily derive from new or historic sources at that site. There are numerous pathways for contamination to migrate on to a property from sources beyond the boundary, and conversely for contamination to migrate off the site on to surrounding land. In the case of groundwater pathways, plumes of contamination may extend for several kilometres, passing beneath numerous properties. Similarly, gas migration can affect properties some distance from the source, as shown by the incidents at Loscoe and Bontddu (see Box 1.1). Migration may be rapid where there are preferential pathways (eg fissures, fractures and underground services) through which contaminants can pass. In other cases, the passage of contamination may be so slow that it can take years or decades before contamination is detectable at a receptor. The fate and behaviour of contamination along the pathway is important in determining the concentration of a contaminant at an exposure point.

2.2.2 Effects of land contamination

The adverse effects of contamination can be seen in terms of harm to a range of receptors, including:

- humans, eg ill health of employees, site visitors, trespassers, contractors working on development sites, neighbours
- the built environment, eg deterioration of building foundations, damage to underground services and impairment of drinking water in mains supplies
- land quality, in terms of a reduction in the capacity of land to act as a reserve of potable water and raw materials, to maintain habitats and biodiversity, to act as a support medium for plant growth and to provide a structural base on which to build.

The likelihood that any such receptors may be harmed by contamination depends on a combination of numerous site-specific factors, such as the concentration and hazardous properties of contamination, the duration of exposure, the exposure mechanism (eg ingestion, inhalation, dermal contact) and the sensitivity of the receptor.

2.3 FINANCIAL RISK

Financial risk stems from an inability, or perceived inability, to forecast the impact of contamination (strictly speaking, the harm that it causes or is perceived to cause) on business costs, income flow and asset value. This is due to a combination of technical, scientific, regulatory, financial and economic uncertainties (see Box 2.2).

Box 2.2 *Examples of uncertainty influencing the financial risk of contamination*

Technical and scientific uncertainties

A principal source of uncertainty lies in the characterisation of a site. Contamination often shows significant temporal and spatial variation. Since only a small fraction of a site is sampled and analysed during even the most comprehensive site investigation (SI), there is always a possibility that an investigation may fail to detect or underestimate the extent of contamination. This problem is compounded by errors in the design and execution of sampling and analytical programmes.

There are also uncertainties relating to the assessment of risk. Many of the parameters used in the analysis of exposure and effect (eg dose response) are poorly understood. New discoveries and developments in science and technology are also continually highlighting new hazards.

Regulatory uncertainties

Much of the statutory environmental legislation which is directly applicable to land contamination in the UK has been introduced since 1990. The principles have yet to be tested fully. Developing scientific knowledge and hardening attitudes to risk may also result in changes to the legislation.

One of the main objectives of regulatory bodies is to enforce legislation in a consistent manner. However, there will always be a degree of variation in decision-making, which reflects differences in individuals' interpretation of information.

Establishing a causal link between contamination and damage can be difficult because of, amongst other factors, the potentially long lag period before the effects of a contaminant release are detected. Multiple changes in site ownership can, in turn, complicate the process of establishing and allocating liability. The aggregate effects of multiple contamination sources on-site and off-site can add to this problem. The exclusion tests under Part IIA EPA 1990 are in their infancy, so it remains to be seen how the exclusion of persons from, and the apportionment of, liability proceed under these and other circumstances.

Financial uncertainties

Environmental liabilities tend to be of the "low probability, high consequence" variety: the chances are that costs will remain within reason, but spectacular cost overruns can occur.

Considerable uncertainty surrounds the cost of being held legally liable. Criminal cases heard at the Crown Court can lead to unlimited fines, whilst civil action can result in unlimited damages awarded against a company. Although rare in the UK, a large number of claims may arise where contamination affects more than one person.

Economic uncertainties

The economy (both at a local, regional or national scale) can greatly influence the financial impact, if any, that contamination has on property prices. Forecasting changes in the economy and its influence on property value is associated with considerable uncertainty.

Degrees of uncertainty vary, as shown in Figure 2.2. In this respect, uncertainty depends on the quantity and quality of information available to evaluate risk. Contamination risk is characterised by significant uncertainty. This can lead to wide differences in the perception of risk.

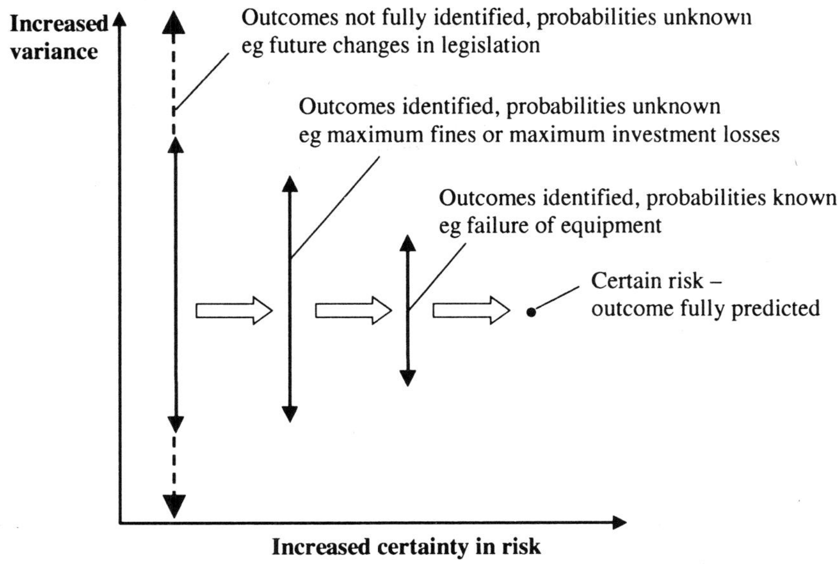

Figure 2.2 *The relationship between risk and uncertainty*

2.3.1 The nature of financial impact

The financial impact caused by contamination comprises:

- legal liabilities (criminal, civil and remediation) – see Section 2.4
- constructive obligations – see Section 2.5
- impairment of physical and financial assets – see Section 2.6

Impact is seen as direct losses (fines, remediation costs etc), or indirect costs caused by uncertainty itself. For example, costs may be incurred as a result of differences in risk perception between a purchaser and a vendor. Uncertainty can also lead indirectly to financial impact as a result of the misallocation of resources for risk management. Provision of non-essential capital will divert resources away from other important business risks. Conversely, uncertainty may mean that a company abstains from an activity which would have generated greater financial benefit than the alternative investments which the company chooses to make.

2.3.2 The likelihood of financial impact

The likelihood that contamination will lead to financial impact depends on, amongst other aspects, the definition of the scope of environmental liability (ie the nature and extent of harm which is actionable under law), the perception of risk amongst stake-holders and property factors.

Scope of environmental liability

In the UK, contaminated land has a statutory definition encompassing specified receptors and levels of harm, as outlined in Table 2.1.

Controlled waters are protected under law (see Section 2.4.2), but without a threshold for harm, ie any entry of polluting matter can in theory trigger liability.

Table 2.1 *Categories of significant harm (excerpt from DETR, 2000)*

Type of receptor	Description of harm to that type of receptor that is to be regarded as significant harm
Human beings	Death, disease, serious injury, genetic mutation, birth defects or the impairment of reproductive functions.
Ecological systems, or living organisms forming part of such a system[1]	Harm which results in an irreversible adverse change, or in some other substantial adverse change, in the functioning of the ecological system within any substantial part of that location; or harm which affects any species of special interest within that location and which endangers the long-term maintenance of the population of that species at that location[2].
Property in the form of crops (including timber); produce grown domestically, or on allotments, for consumption; livestock; other owned or domesticated animals; and wild animals which are the subject of shooting or fishing rights	For crops, a substantial diminution in yield or other substantial loss in their value resulting from death, disease or other physical damage. For domestic pets, death, serious disease or serious physical damage. For other property in this category, a substantial loss in its value resulting from death, disease or other serious physical damage.
Property in the form of buildings	Structural failure, substantial damage or substantial interference with any right of occupation.

[1] Ecological systems or living organisms forming part of such a system are defined in terms of locations which are protected under law.

[2] In the case of a protected location which is a European Site (or a candidate Special Area of Conservation or a potential Special Protection Area), harm is defined as incompatibility with the favourable conservation status of natural habitats at that location or species typically found there.

UK liability law does not cover all forms of environmental damage. Damage to the so-called "un-owned environment" – such as damage to natural ecosystems – does not generally give rise to legal liabilities, except where there are specific statutory regimes in place. For example, it may be an offence under UK law to damage a Site of Special Scientific Interest (SSSI), and contamination harming the ecological system in a SSSI may give rise to remediation requirements under Part IIA. Persons who cause or knowingly permit the entry of pollutants into controlled waters may be required to restore flora and fauna that are dependent on the aquatic environment to their state immediately prior to the pollution. The European Commission's White Paper on *Remedying environmental damage* makes proposals for increasing the scope of liability law to provide for the restoration of particular categories of damaged natural resources.

The UK's approach to the standard of remediation under Part IIA EPA 90 or the planning regime is "suitable for use" whereby risks to human health or the wider environment are managed according to the actual or intended use of the site (see Box 2.1). This contrasts to a multi-functionality approach that is designed to enable any future land use and specifically the most sensitive use. This approach was once favoured in the Netherlands, but has since been modified in view of its high cost to the economy.

Differences in liability regimes between countries can result in the movement of manufacturing facilities to countries with less onerous environmental standards. However, many multinational organisations recognise their moral responsibilities and operate to strict standards wherever their operations are located.

Whether a company will actually be held liable and suffer losses as a result of contamination will be a matter that turns on the facts of each case. Factors influencing this position include:

- the status of an organisation (eg whether under Part IIA EPA 90 the company is a polluter or an owner/occupier)
- nature of the liability (eg strict, retroactive, joint and several)
- the burden of proof (eg on the balance of probabilities or beyond reasonable doubt)
- the availability and effectiveness of defences (eg foreseeability, causation and remoteness, hardship)
- the ability to identify, implicate and transfer some or all of the liability to other persons (as in the case of Part IIA EPA 90)
- the nature and enforceability of any contractual arrangements between organisations (eg warranties, indemnities).

Risk perception

Risk perception often influences the financial impact, if any, that known or suspected contamination has on a business. It can do so by influencing key decisions made by various stakeholders including:

- lenders – their perception will influence the availability and terms of finance
- insurers – their perception will influence the availability and terms of insurance
- investors – their perception can affect share value
- prospective property purchasers – their perception will influence the liquidity of land as an asset and its value in the open market
- regulators – the views of the local planning authorities (LPAs) will influence the need for, and costs of, remediation on development
- general public – the public's perception of a company can influence the revenue that it makes from selling goods or services

Petts (1994) describes stigma as a negative perception of an area as a result of an environmental problem that may encompass feelings of disruption, loss of amenity, danger, ugliness or loss of value. It may be considered as the value impact of environmentally-related uncertainties; a negative intangible likely to be caused by one or a combination of the following factors:

- the fear of hidden clean-up costs, including the uncertainty of cost over-runs, possible market resistance after treatment, the possibility of unforeseen contamination and the risk of failure of remediation
- the cost of compensation to a potential purchaser for dealing with the problem
- the perceived higher risk of future liability compared to an uncontaminated site
- the lack of mortgageability, including the inability to secure financing for the sale or development of the site
- the risk of increased future liabilities as a result of future changes in legislation or remedial standards
- a reduced range of alternative uses of the site, reducing its future marketability.

Property factors

Property factors include the effects of location and supply and demand on the value of a property. The location of a property is often the over-riding factor in determining its demand, and thus its market value. It appears to extend to contaminated sites, with a strong demand often counteracting the effects of contamination on property value (Smith, 1999). Geographical location and demographical distribution may, therefore, have a role to play in determining whether a contaminated site will be redeemed by favourable market conditions, and may reduce the need for protective measures.

2.3.3 Financial risk exposure

A company's exposure to financial risk depends on its particular involvement and interest in a site, as shown in Table 2.2.

In general terms, the closer an organisation comes to having an interest in, or control over, the occupation of land, the greater the exposure to front-line, contamination-related financial risks (see Figure 2.3). For example, developers, landowners, tenants and homeowners have a primary interest in land and may be held legally liable for contamination. Contractors working on a contaminated site may also be exposed to these risks, although this depends largely on the terms of their contract and their role in a project.

Lenders and investors, which under normal circumstances only have a secondary interest in a site, are generally not directly exposed to the financial risks of contamination. Instead, their main exposure consists of impairment of financial assets including mortgage repayments, capitalised interest, security on a loan, initial equity and return on an investment. Such losses can arise if contamination reduces a borrower's share price or forces the company to default on a loan or other secured finance.

Land contamination liabilities tend to remain hidden, only to be discovered in the course of asset or share transactions, mergers and acquisitions (M&A), or during development projects. Occasionally losses arise unexpectedly through the latent effects of historic contamination, or as a result of releases of new contamination (see Box 1.1).

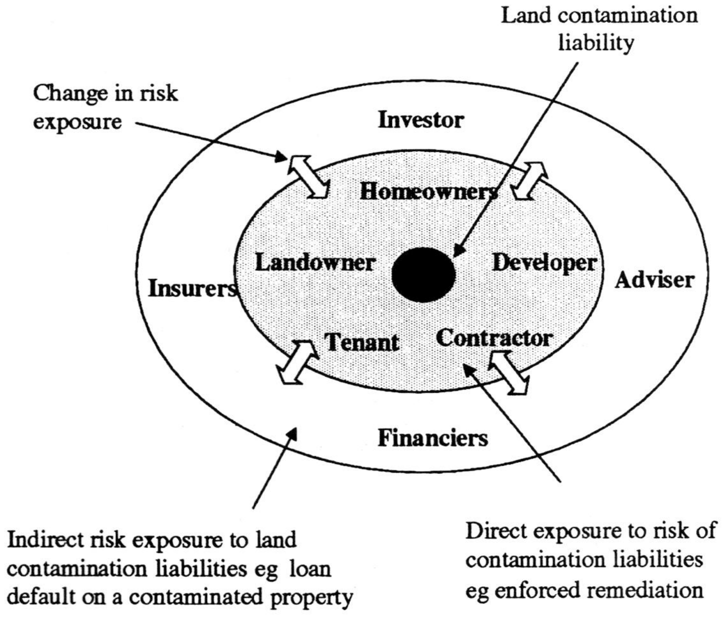

Figure 2.3 *Exposure of organisations to legal liabilities*

Liabilities may also materialise as a result of the local authorities' duty to inspect their areas for contaminated land – see Section 2.4. In announcing the implementation of the contaminated land regime, the Government issued a challenge to all major businesses to develop and publicise their own company strategies for finding and dealing with the sites for which they are responsible.

Table 2.2 *The nature of financial risk exposure for specific activities*

Financial risk exposure / **Activities**	**Homeownership and occupancy**	**Land ownership (potentially polluting)**	**Land ownership (non-polluting)**	**Leasing of property to third parties**	**Land occupancy**	**Land purchase**[1]	**Land sale**[1]	**Land development**	**Provision of finance**	**Investment**	**Contracting (including remediation)**	**Provision of professional advice**
Criminal liabilities												
Environmental offences		✓✓			✓✓			✓✓	✓[2]		✓✓	
Health and safety offences		✓✓	✓	✓	✓✓			✓✓	✓[2]		✓✓	
Civil liabilities												
Third party liabilities[3]	✓	✓✓	✓	✓	✓✓			✓✓	✓[2]		✓✓	
Breach of professional duty of care								✓			✓✓	✓✓
Breach of contractual arrangements[4]				✓	✓			✓			✓✓	✓✓
Remediation liabilities	✓	✓✓	✓✓	✓	✓			✓✓	✓[2]		✓✓	
Impairment of assets												
Depreciation in land value	✓✓	✓✓	✓✓	✓✓	✓✓			✓✓	✓[2]			
Restrictions on land use	✓	✓✓	✓✓	✓✓	✓✓			✓✓	✓[2]			
Project[5] disruptions		✓✓	✓		✓	✓✓	✓✓	✓✓			✓✓	
Reduced liquidity of asset		✓✓	✓✓	✓	✓✓			✓✓				
Loss of investment (equity)								✓[6]		✓✓		
Default on loan or other secured finance									✓✓			
Damage to corporate/brand reputation		✓✓	✓	✓				✓✓	✓✓	✓✓	✓✓	✓✓

The number of ticks indicates the relative risk to which an organisation is exposed through undertaking the particular activity.

1. Contamination can disrupt a transaction or in the worst case cause it to be aborted. On completion of the transaction, the purchaser may acquire any or all of the liabilities listed under the land ownership columns. The vendor may also still retain a proportion of these liabilities. The allocation of risk is project-specific and dependent on, amongst other factors, the contractual arrangements made between the vendor and purchaser.
2. Only in the event that a borrower defaults on a loan and the financier takes possession of the property as security on the loan.
3. In respect of bodily injury and property damage.
4. For example, breach of terms of lease, service contract or warranty.
5. Projects may take various forms including remediation, land development, buying a property or making improvements to reduce the risk of pollution on an operating facility.
6. Only applicable to developer-investors who invest in a development project (as opposed to developer-traders who do not take an equity investment in the project).

2.4 LEGAL LIABILITIES

2.4.1 Legal framework

Land contamination is regulated under several regimes, including environmental protection, pollution prevention and control, waste management, planning and development control, and health and safety. Depending on the legislation, liability can be strict, retroactive, joint and several or may take another form as under Part IIA, under which liability may be shifted from one responsible person to another.

Within the UK legislative framework, there is a distinction between statutory controls on authorised activities that are causing or may lead to contamination, statutory liability for contamination from unauthorised activities, and legislation that seeks to control or remove residual contamination resulting from historic incidents.

Control of contamination from existing activities

Contamination releases from current activities normally fall within the following licensing/permitting regimes:

- authorisations under Groundwater Regulations 1998
- Waste Management Licensing under Part II EPA 1990
- Integrated Pollution Control (IPC) authorisations under Part I EPA 1990 (to be replaced by permits under Pollution Prevention and Control (PPC) Act 1999)
- authorisations under Radioactive Substances Act 1993.

The responsible body (normally the Environment Agency or SEPA) has powers to issue prohibition notices where there is an imminent probability of serious environmental damage, suspend or revoke environmental licences, vary licence conditions, seek injunctions or carry out remedial works. Where the responsible body has carried out remedial works, it generally seeks to recover the full costs incurred from those responsible.

Liability for contamination from unauthorised activities

Where there is an unauthorised release of a pollutant, there are also provisions to serve notices to prevent water pollution and to remove waste (see Section 2.4.2). These are served at the discretion of the responsible body where it is necessary to prevent future pollution or to undertake works to mitigate the effects of polluting activities.

Remediation of historic contamination

The remediation of historic land contamination will continue to be managed through the planning regime. Planning Policy Guidance Note 23, *Planning and pollution control* (currently under revision), reiterates earlier guidance that ground contamination is a "material planning consideration". This means that local planning authorities (LPA) must take contamination into account when considering a planning application for a site that is suspected or known to be contaminated. In such a case, the LPA may be required to consult authorities such as the Environment Agency. The LPA may require a developer to investigate, assess or remediate the contamination, for example, as a condition to being granted planning permission. The standard of remediation is suitability for the proposed use. This is a risk-based approach that considers the risks posed by contamination on the site to various receptors including, but not limited to, humans, buildings and the environment.

Part IIA EPA 1990 (implemented via the Contaminated Land (England) Regulations, 2000 and statutory guidance [DETR, 2000]), "the contaminated land regime", complements the planning process by dealing with those sites which, in their current or likely use not requiring planning permission, present unacceptable risks to human health or the wider environment, and which are not being actively redeveloped. This regime replaced the sections of the statutory nuisance system in respect of land "in a contaminated state". LAs or, the Environment Agency and SEPA in the case of "special sites", have powers under Part IIA to require appropriate persons to carry out remediation, or undertake works themselves and recover their reasonable costs from the appropriate person(s) (see Section 2.4.2).

In certain circumstances, the Environment Agency/SEPA can serve a Works Notice on the person or persons who caused or knowingly permitted pollution of controlled waters (s161A WRA 1991 implemented via the Anti-Pollution Works Regulations 1999 for England and Wales). The Works Notice requires the person who caused or knowingly permitted the pollution to remedy it, and if reasonably practicable, restore flora and fauna associated with the polluted waters to their state immediately before the pollution (see Section 2.4.2).

Health and safety

A contaminated site can also fall under the jurisdiction of the Health and Safety Executive (HSE). This body has the primary responsibility for enforcing health and safety legislation in England and Wales.

Private law

An increasingly important source of financial risk comprises private actions brought in tort by third parties whose interests have been damaged (eg suffered bodily injury or property damage), by parties to a contract or by environmental interest groups. The key torts in relation to land contamination are negligence, the rule in *Rylands* v *Fletcher* and private nuisance (see Box 2.2).

The criminal and civil liabilities that may arise through statutory and private law are discussed in Sections 2.4.2–2.4.4.

Box 2.2 *Principal torts relevant to land contamination (Laidler et al, unpublished)*

Private nuisance: it is a civil wrong giving rise to a claim to interfere with the reasonable use and enjoyment of another's land or of some right over or in connection with it. Liability arises where the damage caused by the nuisance was reasonably foreseeable at the time of the relevant act or omission.

Negligence: the breach of duty of care owed by one person to another resulting in damage to the other is a civil wrong giving rise to a claim in damages. Actions for negligence in relation to environmental matters are generally difficult to sustain because of the difficulties which can arise in showing the existence of a duty, causation and a failure to meet the standard of care necessary to discharge that duty.

Rule in Rylands v ***Fletcher (1866)***: in essence the rule states that where a person brings on to his land for his own purposes anything likely to do mischief, if it escapes he is liable for the reasonably foreseeable damage which is the natural consequence of its escape from the land, regardless of the absence of fault in the cause of the escape. The rule does not apply where the use of the land involved is natural. The ruling in *Cambridge Water* v *Eastern Counties Leather* established the problem in foreseeing damage from historic contamination, but the foreseeability of the consequences from more recent contamination.

2.4.2 Remediation liabilities

A form of civil liability is created by a number of statutory provisions which allow authorities to:

- remediate contamination (s78P EPA 1990)
- take measures in relation to actual or likely water pollution (s161 WRA 1991)
- remedy damage caused by an IPC process (s27 EPA 1990)
- require the removal of illegally deposited (ie fly-tipped) waste (s59 EPA, 1990).

Part IIA of EPA 1990 allows authorities, in certain cases, to secure the cost of remediating contaminated land by way of a charging order (a statutory land charge) on premises comprising the contaminated land in question.

Contaminated land (Part IIA EPA 1990)

Contaminated land is defined by Part IIA as "any land which appears to the local authority in whose area it is situated to be in such a condition, by reason of substances in, on or under the land, that:

- significant harm is being caused or there is a significant possibility of such harm being caused or
- pollution of controlled waters is being, or is likely to be caused."

The word "significant" is described by criteria in statutory guidance (DETR Circular 2/2000) in relation to receptors, types of harm and possibility of harm (see Section 2.2). There is no threshold requirement of harm for controlled waters but the government does not intend to require remediation where harm is trivial.

If a local authority decides, on the balance of probabilities, that there may be a significant pollutant linkage (ie a contaminant, one of the receptors specified in the statutory guidance and a pathway linking the contaminant and receptor) the land will be designated contaminated land. A subset of contaminated land will be classified as special sites. These include sites at which groundwater in certain aquifers (as specified in the Contaminated Land (England) Regulations, 2000, SI 2000 No 227) are contaminated by compounds defined in List I of the EC Groundwater Directive, where potable groundwater is threatened or watercourses would fail criteria for classification under regulations made under the WRA 91, or where oil was refined, explosives were manufactured or the site is occupied for Ministry of Defence purposes. The Environment Agency and SEPA are the enforcing authorities for special sites.

The regime imposes strict and retroactive liability for remediation of contaminated land on two classes of appropriate persons. Primary liability rests with Class A appropriate persons who caused or knowingly permitted the presence of contaminants on land. If an enforcing authority fails to "find" a Class A appropriate person after a "reasonable inquiry", the current owner or occupier of the contaminated land becomes liable as a Class B appropriate person. An organisation or other person can knowingly permit a substance to be present if it knows of the substance's existence, has the power to prevent its existence or to remove it and has a reasonable opportunity to do so.

Remediation requirements include assessment, treatment, measures to break the pollutant linkage and/or monitoring. In all cases, remediation must be practicable, effective and durable, and its cost must be reasonable with regard to the cost of remediation and the seriousness of the contamination. There is no monetary limit on liability for a Class A appropriate person, whilst a Class B person's liability is restricted to remediating the land that they own or occupy. The remediation standard under Part IIA is suitability for current and likely use. A likely use is one that does not require planning permission and is not speculative. An exception exists when an appropriate person causes land to be contaminated through a breach of an environmental licence or permit. In such cases, the contamination must generally be completely removed.

Where more than one appropriate person is liable for remediating a significant pollutant linkage, the enforcing authority must apply one or more tests to exclude some of the appropriate persons from the liability group. The authority must cease applying the tests if doing so would exclude all appropriate persons.

There are two sets of exclusion tests: one for Class A persons and one for Class B persons. Both sets of tests are applied in the order in which they appear in the statutory guidance, but must not be applied to exclude all persons in the liability group. **The exclusion tests for Class A appropriate persons are:**

- excluded activities
- payments made for remediation
- sold with information
- changes to substances
- escaped substances
- introduction of pathways or receptors.

The statutory guidance contains criteria to apportion remediation costs between appropriate persons who remain in a liability group after the exclusion tests have been applied. It also sets out criteria to attribute costs between more than one liability group when more than one significant pollutant linkage exists at a site. The guidance also contains hardship criteria to which an enforcing authority must have regard.

When Part IIA applies due to the existence of contaminated land, the enforcing authority notifies each appropriate person in the liability group for each significant pollutant linkage. The Government's intention under Part IIA is that enforcing authorities and appropriate persons should, wherever possible, arrive at a voluntary agreement for the remediation of contaminated land, without resorting to enforcement measures. In cases where voluntary action is not carried out, the local authority must issue a remediation notice stating the work that must be carried out. There is a three-month period for voluntary action after the enforcing authority has issued notifications, before it may issue a remediation notice.

If an enforcing authority undertakes remediation of a contaminated site, it may place a legal charge on the site for its reasonable costs if a Class A person owns the land. The legislation does not specify whether such a charge has priority over a prior mortgage and other interests but analogous case law exists to imply that it will do so.

Enforcing authorities must compile public registers to include remediation notices, appeals against them, designation of special sites and remediation statements (ie statements by appropriate persons detailing remediation actions to be conducted at contaminated sites, when the actions take place after notification has been served but a remediation notice has not been served). As such, the location of contaminated sites will become public knowledge, although the register will not list the contaminated sites other than special sites and sites at which the enforcing authority is precluded from serving a remediation notice. There is no mechanism for removing information from the public registers. Most persons, including homeowners, can become liable for contaminated land under certain circumstances. These circumstances are discussed in Chapters 5–9.

Pollution of controlled waters

Sections 161A to 161D WRA 1991 (implemented via the Anti Pollution Works Regulations 1999) authorises the Environment Agency to serve Works Notices on person(s) who caused or knowingly permitted water pollution to arise, in order that that it can be remedied or prevented. Sections 46A to 46D of the Control of Pollution Act 1974 applies to Scotland. Owners and occupiers of land will not be liable in these capacities merely for owning or occupying land. There are no statutory provisions for, or guidance on, allocation and apportionment of liability. Joint and several liability may arise.

IPC (to be replaced by PPC)

The Environment Agency and SEPA have the power under section 27 EPA 1990 to take action to remedy harm that has been caused due to the breach of an authorisation which has been granted under the Integrated Pollution Control regime (IPC).

In 1999, Parliament transposed the EC Directive on Pollution Prevention and Control into UK law. The legislation is being phased in by industry sector, ending in 2007 for existing installations. Under both the IPC and PPC regimes, the Environment Agency/ SEPA may serve an enforcement notice on a person who breaches the terms and conditions of an authorisation (known as a permit under PPC) or who operates a designated polluting process (or, under PPC, an installation such as a manufacturing facility) without an authorisation.

In addition, persons applying for a permit under PPC will be required to assess the land on which the permitted installation is located and prepare a report on the condition of the land. When the installation ceases operation, the land must be reassessed and any deterioration in quality must be rectified so that the site of operation is returned to "a satisfactory state".

Waste management licensing

If the holder of a waste management licence breaches the terms and conditions of the licence and the breach results in contamination, the operator may be required to remediate the contamination. If the operator causes contamination other than by a breach of the terms and conditions of the licence or other than in respect of any action authorised by the terms and conditions of the licence, the operator may be required to remediate the contamination under Part IIA EPA 1990. Operators of sites that are covered by waste management licences are not permitted to surrender the licences until they have ensured that the site does not pose harm to people or threaten to pollute the environment.

If a company deposits controlled waste (ie industrial, commercial or household waste) on a site for which a waste management licence has not been issued, the company may be required to eliminate or mitigate the consequences of the deposit under section 59 EPA 1990. The legislation does not provide a standard to which the remediation must be conducted.

Cost of remediation

The costs of remediation are highly variable and dependent on site-specific characteristics such as the nature and extent of contamination, the ground conditions and the type of remedial strategy selected. Indicative ranges of costs for various remedial techniques are presented in Table 2.3.

Table 2.3 *Indicative remediation costs (Nathanail, 2000)*

Treatment technique	Indicative unit prices
Engineering capping	£15–£30/ m^2
Excavation and disposal to landfill	£50/m^3
Encapsulation (shallow cut-off wall)	£40–£60/ m^2
Encapsulation (deep cut-off wall)	£70–£120/ m^2
"Typical" landfill gas control system	£200 000
"Typical" grout curtain/vent trench	£220 000
Bioremediation	£35–£45/ tonne
Vitrification	£40/ tonne
In-situ vitrification (5t/hr)	£150–£215/ tonne
Incineration (special wastes)	£750–£1000+/ tonne
Dechlorination	£100–£300/ tonne
Soil vapour extraction	£40–£60/m^3 vadose zone
Soil washing	£30–£35/ tonne
Enhanced thermal conduction	£35–£45/m^3
In situ chemical oxidation	£40–£80/m^3
Pump and treat	£20–£30/m^3
Free product recovery	£10–£20/m^3 vadose zone
Air sparge	£45–£55/m^3 groundwater
Solvent extraction and incineration	£400/tonne
Thermal desorption (including excavation and pre-treatment)	£35–£150/ tonne

2.4.3 Civil liabilities

Civil liability generally relates to breaches of contract and the law of tort. Unlike criminal law, which requires the prosecution to prove beyond reasonable doubt that the defendant is guilty, the standard of proof in civil cases generally rests on the balance of probabilities.

Liability in contract for contamination may arise under a contractual agreement such as an indemnity, warranty, or a landlord or tenant's covenant in a lease. Remedies for breach of contract include damages and/or rescission. Parties to a contract may insert a provision stating the liquidated damages applicable to a breach of contract. To be enforceable, this amount must represent a genuine pre-estimate of the actual loss that would be suffered by the innocent parties and should not constitute a financial penalty.

Damages in tort may be awarded due to death or injury; disease caused by exposure to a hazardous substance; disease or injury to livestock; physical damage to property and other unreasonable interference with property rights. Remedies for harm also include injunctions which either require that certain actions be carried out, or prohibit actions that may lead to a described state of affairs. A final injunction will be granted only after a full trial, but a court may grant an injunction pending the full trial (interlocutory injunction) if the balance of convenience is in favour of doing so.

The amount of compensation awarded by the courts is calculated as the sum of money which (so far as money can do so) puts the plaintiff back into the position he would have enjoyed but for the defendant's wrongdoing (Graham, 1995). The damages that result from civil pollution liability claims can be considerably greater than the typical fines imposed for environmental offences. The coincidental costs of civil litigation, including legal costs, wasted management resources and delays to projects, can also be significant. Examples of tort actions involving environmental liabilities are presented in Table 2.4.

Table 2.4 *Examples of civil action involving land contamination*

Civil wrong	Example of damages awarded
Death or personal injury	*Hancock* v *JW Roberts Ltd (1996).* £65 000 and £50 000 damages awarded following death and personal injuries respectively, due to exposure to asbestos dust in vicinity of factory. Evidence indicated that the company had known of the harmful effects of asbestos, but had taken no action to protect persons exposed to the asbestos.
Nuisance	*Cambridge Water Company* v *Eastern Counties Leather plc (1993).* No damages awarded for contamination of public water supply borehole because the consequence of chemical spillages at the tannery's property could not have been reasonably foreseen at the time of the spillage.
Negligence	*Urban Regeneration Agency* v *Mott MacDonald Ltd (1998).* £18.5 m damages awarded for negligent advice provided by a developer's consultants. An appeal was made and the case was settled out of court for an undisclosed sum of money before the appeal was heard.
Breach of statutory duty	*Blue Circle Industries* v *Ministry of Defence (1996).* £6 m damages awarded for compensation for breach of statutory duty under the Nuclear Installations Act 1965. Damage included economic loss arising from a sale of the property falling through because of the discharge of radioactive material onto Blue Circle's property.

2.4.4 Criminal liability

Environmental and related legislation creates a number of offences for environmentally damaging, or potentially damaging activities. Where a criminal offence has been committed, the enforcing body (normally the Environment Agency or, in some instances, the Procurator Fiscal on behalf of SEPA) will consider the following course of action in addition to any other enforcement action (Environment Agency, 1998).

1. Issuing a warning: a written notification that, in the Environment Agency's opinion, an offence has been committed. It is recorded and may be referred to in subsequent proceedings.

2. Administering a caution: a written acceptance by an offender that he has committed an offence. It may only be used where a prosecution could properly have been brought. It will be brought to the court's attention if the offender is convicted of a subsequent offence.

3. Instituting a prosecution. Most environmental offences can be tried "either way", that is they can be heard in the magistrates' court or the Crown Court in England. Depending on the court and offence, a range of penalties may apply. In the magistrates' court, the maximum penalty on conviction is a fine of £20 000 and/or six months imprisonment, whereas cases taken to the Crown Court can carry unlimited fines and up to two years' imprisonment. A convicted defendant may also have to pay for the costs of the prosecution, including investigation costs incurred by the regulator.

Where a company is involved it is usual practice to prosecute the company where the offence resulted from the company's activities. The enforcing authority also considers any part played in the offence by the officers of the company, including the directors, managers and the company secretary. In appropriate cases, the authority will prosecute a director or officer and consider seeking disqualification of directors under the Companies Act.

Offences

There are no specific offences for contaminating land as such, but a number of circumstances exist where acts or omissions can lead to the commission of criminal offences. These include the following.

1. Breach of conditions attached to a licence or authorisation.
2. Failure to comply with a remediation notice.

If the land is an industrial, trade or business premise, an enforcing authority may impose a fine of up to £20 000 and an additional fine of up to £2 000 per day under Part IIA until remediation begins. There are no defences to liability, except for reasonable excuse, but the recipient of a remediation notice may challenge the notice on one or more of twenty four grounds, including the grounds that the notice is defective, the land is not contaminated or the remediation requirements are not reasonable.

3. Failure to comply with Works Notices.
4. Polluting controlled waters.

The main water pollution offences are contained within s85 of Water Resources Act 1991. Subject to defences, a person commits an offence by causing or knowingly permitting any poisonous, noxious or polluting matter or any solid matter to enter any controlled waters.

Examples of penalties arising from environmental offences are presented in Table 2.5.

Table 2.5 *Examples of penalties arising from environmental offences*

Offence	Legislation	Example
Pollution of controlled waters	WRA 1991 Control of Pollution Act 1974 (Scotland)	***NRA* v *Shell* (1990):** £1 m fine
Deposit or handling of waste without (or in breach of) waste management licence	EPA 1990 Part II	*Lanstar (1999):* £40 000 fine for breach of licence *EQM Construction Ltd (1998):* £21 000 fine for illegal waste disposal
Disposing of controlled waste in a manner likely to cause harm to human health	Section 33(1)(C) EPA 1990	*Thames Water* (1998): £200 000 for illegal disposal of controlled waste, £50 000 for pollution of controlled waters plus costs to the Environment Agency of £12 847
Pollution of controlled waters	WRA 1991	
Failing to ensure that premises are safe	Health and Safety at Work etc. Act 1974	*Rollco Screw and Rivet Co. (1998):* Brown asbestos on site presented a danger to employees

2.5 CONSTRUCTIVE OBLIGATIONS

A company may be under a constructive obligation to undertake remediation works irrespective of whether or not it is legally liable. This situation occurs where an organisation has created a valid expectation in its customers, suppliers, employees and other stakeholders that it will take on and fulfil its clean-up responsibilities by virtue of its published policies or established past practices. In such cases, the company is able to recognise a balance sheet provision for the remediation, or put aside provisions to undertake the works (ASB, 1998).

2.6 ASSET IMPAIRMENT

2.6.1 Damage to physical assets

The presence of known or suspected land contamination may adversely affect the value of a property asset in the open market. The highest value which may be realised for such an asset is that which a buyer is willing to pay, subject to town planning controls, after deduction of the lowest cost of remedying the effects of the contamination. Examples of how the asset value may be affected by contamination include:

- decreased liquidity – less people may be willing to purchase the property, and hence the realised value may be reduced. Property assets are generally less liquid than shares which can be readily resold on the stock market.
- increased transaction costs – the need for due diligence may reduce the net income to the business
- inability to generate income as the asset is unavailable
- increased asset realisation period – it may take longer to realise the value of an asset as the contractual arrangements and due diligence relating to realisation may be more lengthy and complex
- liabilities upon asset disposal – the disposal or the valuation of an asset may give rise to contingent liabilities which need to be provisioned at the point of asset valuation.

The main cost factors to be considered when valuing a property which may be contaminated fall under the broad headings of direct and indirect costs, and property factors.

1. Direct costs

Direct costs are principally concerned with the costs of remediation and monitoring, including site investigation and consultancy advice. Void costs (where occupied buildings may need to be vacated) may also be significant. In a let property this will mean the incorporation of rental void period. Direct costs may also include penalties for non-compliance with legal requirements.

Where the costs of contamination exceed the market value of a property, a negative property value will result. When a property comprises part of the assets of a company, a valuer is under a duty to ensure that such negative values are reported separately in the valuation certificate, and not set off against positive values on other properties.

2. Indirect costs

As a result of stigma, the presence of known or suspected contamination can have implications extending beyond the costs of simply remediating the land, and could be expected to impact upon the yield rate, the mortgage terms available and the future anticipated appreciation or depreciation of the property concerned. There will always be potential purchasers that will not consider a former contaminated site for their own use or development, even if all traces of contaminants have been removed. The marketability of a site with low levels of residual contamination is therefore likely to be reduced even further.

2.6.2 Damage to financial assets

The presence of known or suspected land contamination may also adversely affect the financial asset in terms of:

- default on loans and other secured finance – lenders may suffer loss of principal and capitalised interest
- investment – investors may not recover the value of their initial investment and fail to obtain the forecasted rate of return on investment
- corporate/brand reputation – just as stigma can constitute an unquantifiable negative impact on the value of land, so harm to reputation can impact on share price or other measures of an organisation's value.

2.7 BUSINESS IMPACT

The impact of contamination is typically seen as extra costs to a business in terms of increased capital expenditure, reduced revenue and impaired asset value. The impact can be felt throughout a company. At a group level, the effect may be felt in terms of a reduction in share price or through a prosecution of a director or officer. The company may also be seen as a less attractive investment. In extreme circumstances, the impact may be so great that a company is forced into liquidation or insolvency. At an operational level the effects of contamination can cause budgets to be exceeded, disrupt income flow and reduce staff morale.

3 Financial risk management

3.1 PRINCIPLES OF FINANCIAL RISK MANAGEMENT

Financial risk management encompasses policies, processes and practices applied by businesses to protect their economic interests. Companies seek, where possible, to minimise their risk of suffering financial losses. This may be achieved by abstaining from activities or projects that expose the company to a particular risk (**risk avoidance**). Alternatively, the risk may be transferred to another organisation (**controlled risk transfer**). A third option involves retaining and controlling the risk using techniques to reduce the probability of a loss materialising (**loss prevention**) or minimise the impact of a loss when it occurs (**loss mitigation**). Collectively, these techniques reduce a company's exposure to financial risk. However, risk cannot be eliminated and funds may, therefore, be needed to finance the losses associated with the residual risk (**risk financing**). These funds may be derived from a company's internal resources (**retained risk financing**), external sources (**financial risk transfer**) or a blend of the two (**composite risk financing**).

Organisations may also attempt to reduce the uncertainty of their financial risk exposure. This can be achieved by undertaking risk assessment and by adopting risk management techniques which provide greater certainty in the level of exposure (see Section 3.1.1).

The principles of managing financial risk (illustrated by crosses) and uncertainty (illustrated by shaded squares surrounding the crosses) are illustrated in Figure 3.1.

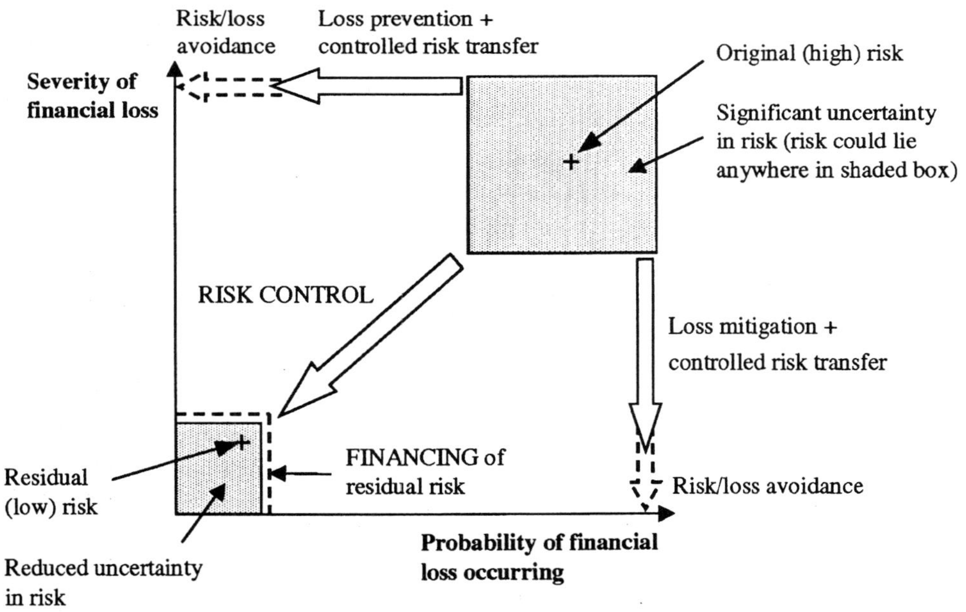

NB The size of the boxes represents the level of uncertainty in the risk – the larger the boxes, the greater the level of uncertainty.

Figure 3.1 *Principles of financial risk management*

3.1.1 Risk management processes

Risk management often seeks to alter an organisation's risk exposure by changing one type of risk for another. This in turn changes the nature of uncertainty (technical, legal and financial) of the risk, which, depending upon an organisation's perception, may provide the company with greater confidence in its risk exposure. For example, an organisation may be more comfortable with the legal uncertainty associated with the wording of an insurance product than the financial uncertainty associated with the creditworthiness of a warrantor. In this respect, an organisation's risk management strategy will depend on its perception of risk and uncertainty.

If a particular risk is not acceptable to a company then, depending on the costs, additional or alternative risk management steps may be sought in order to reduce risk and uncertainty. Several such changes may be necessary to achieve acceptable levels of risk and certainty, as illustrated in Figure 3.2.

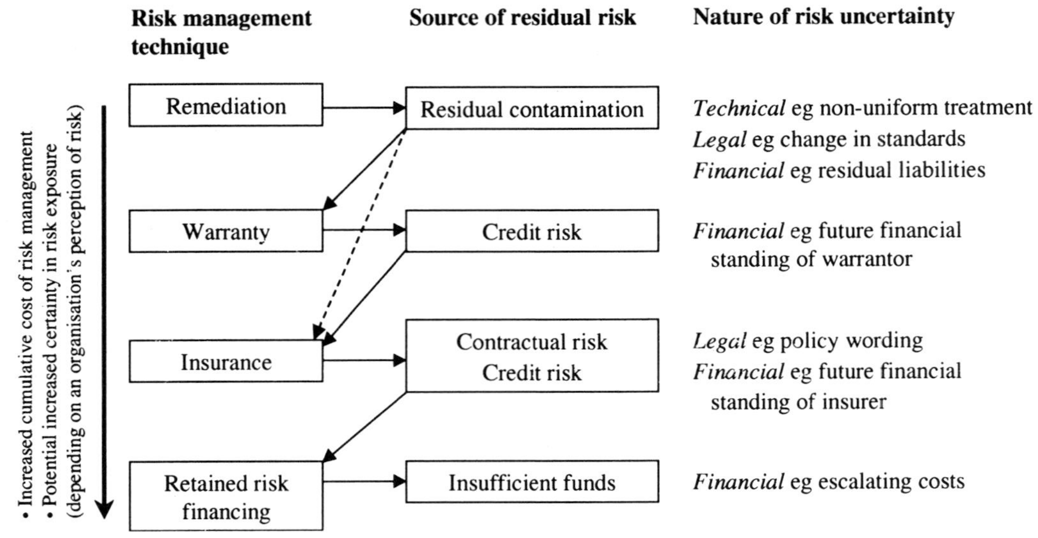

Figure 3.2 *Effects of risk management techniques on risk and uncertainty*

3.1.2 Economics of risk management

In practice a balance needs to be struck between the business benefits (eg risk reduction and increased certainty in the risk exposure) that management techniques generate and the corresponding costs of implementation.

In environmental management, a relatively small additional cost can often result in a significant reduction in financial risk (shown by a high benefit:cost ratio). For example, a soil treatment programme costing a few thousand pounds can minimise the risk of groundwater pollution for which the remediation costs could run into several millions of pounds. Obtaining more information on a risk can also generate value by reducing uncertainty (see Box 3.1). However, the costs of obtaining more information or implementing additional risk management often quickly outweigh the benefits of achieving further reduction in risk or uncertainty (the law of diminishing returns). In such cases, any additional expenditure on risk management may not be economically justified.

The course of implementing techniques, either singularly or in combination, can also influence the economics of risk management. For example, in Figure 3.2 it may be more cost-effective to seek insurance against the effects of residual contamination (as shown by the dotted line) than to obtain a remediation warranty backed by insurance.

Box 3.1 *Example of the value of undertaking more comprehensive risk analysis*

> Harpurhey reservoirs lie in North Manchester, about two miles from the City Centre. Owned by Manchester City Council, it was known that the reservoirs were contaminated. The council procured various site investigations to consider the remedial works and the costs to clean up the land to make it safe for use as an urban nature reserve. Budget costs for remediation from previous studies were in the order of £1.2 million. A more comprehensive design study was commissioned to establish the detailed costs. This study revealed that an alternative approach could remediate the site to a standard fit for its end use for £330 000 – a saving of £870 000 on the original budget. Further savings were identified through the use of materials from the construction works of a nearby stadium. The additional engineering advice cost £25 000 – more than paying for itself.

3.1.3 Proactive versus reactive management

There has been some debate, both from a legal and commercial perspective, on how proactive organisations should be in managing their contamination-related financial risk exposures. This stems from a view shared by some that contamination losses are a minor issue in relation to other business risks. There is also a possibility that an owner/ occupier in a Class B liability group under Part IIA (see Section 2.4.2) may elevate itself into a Class A knowing permitter liability category if it acquires knowledge of a substance's existence, has the power to prevent its existence or to remove it, and has a reasonable opportunity to do so. Arguments against pro-activity are, however, countered by a number of reasons, including:

- companies are not only exposed to Part IIA liabilities – private law actions can be equally, or more damaging. A proactive approach to environmental risk management can help prevent escalation of such risks
- recent cases (see Box 1.1) have shown the potentially vast sums of money that can be incurred as a result of land contamination incidents
- acquiring information on the risks of contamination can enable more informed decision-making on the need for, and financial benefits of, various risk management options. This helps to ensure that the company is appropriately protected against financial losses, and can be used to plan acquisition, disinvestment, redevelopment and other business strategies. Awareness of existing risks can also be used to minimise problems in the future, for example, by ensuring that work on the site does not exacerbate historic contamination
- the discovery of unknown contamination during a transaction can have serious implications for the success of the deal, particularly when it needs to be completed quickly. In the worst case, the transaction may be aborted or the vendor may be forced to concede onerous indemnities and warranties to the purchaser or settle for a reduced sale price. Obtaining information in advance of a transaction can help to avoid the "deal breakers" and support the negotiation process
- a policy of transparency and partnership with the regulator may provide a smoother and quicker passage through the regulatory system. A demonstrable proactive risk management strategy, based on risk assessment, will also help to meet the internal control requirements of the Code (see Section 1.2.4), as well as providing a basis on which to respond to any stakeholders' concerns regarding contamination risk
- evidence of a proactive approach to risk management will place the company in a better position to counter any malicious claims of poor environmental management
- minimises the risk of a blighting effect which the placing of a property on a register under Part IIA may create.

3.2 MANAGEMENT TECHNIQUES

Techniques to manage financial risks can be divided into two broad categories, control and financing, according to their mode of action (see Figure 3.3).

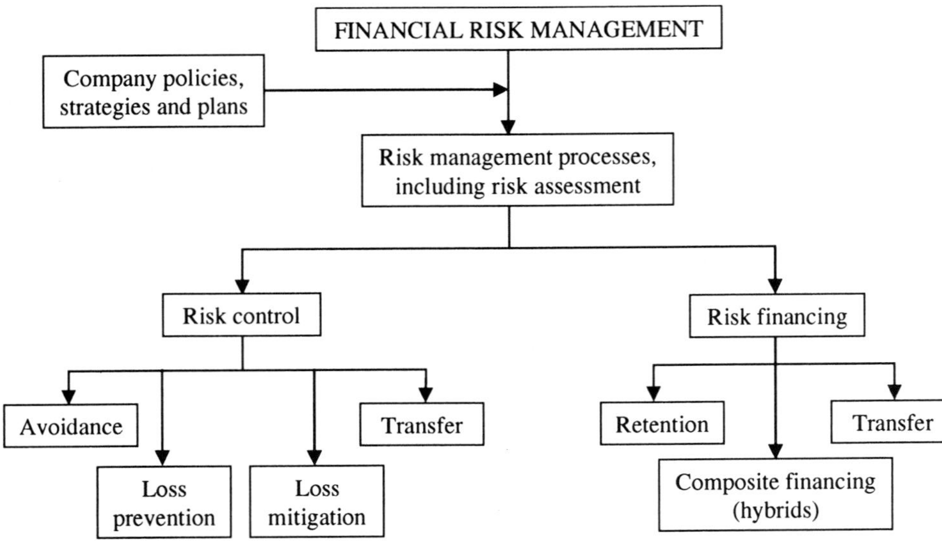

Figure 3.3 *Categories of financial risk management techniques*

3.3 RISK CONTROL

Risk control methods reduce a company's exposure to risk by helping it to avoid a risk, reducing the probability that a loss occurs, lessening the impact if a loss occurs, or reducing the undesirable effects of a risk. Techniques can be categorised into risk and loss avoidance, loss prevention, loss mitigation and certain forms of risk transfer.

3.3.1 Risk and loss avoidance

Risk avoidance (and thereby loss avoidance) is achieved by abstaining from the activity or property which generates the risk of loss. Whilst the process ensures that a company will not experience the losses or the uncertainty that the risk may create, it also means that the benefits associated with the activity may be lost. This often makes avoidance an unacceptable option. Similarly, avoidance is unsuitable where the risk forms an integral part of a business. It is clear that the risks associated with known or suspected contamination have contributed to companies avoiding the purchase or development of certain sites. However, in most cases it is not practicable to avoid the risks altogether because land ownership and occupancy forms an integral part of most businesses' capital structure. It is also virtually impossible to exclude contaminated sites from a loan book or an investment portfolio since any site – greenfield or brownfield – may be contaminated to some degree.

3.3.2 Loss prevention

Loss prevention techniques are designed to reduce the number of losses or to eliminate them entirely. The techniques aim to modify the hazard, alter the environment in which the hazard exists, or intervene in the processes whereby hazard and environment interact. Examples of loss prevention measures applied to land contamination risks are presented in Table 3.1.

Table 3.1 *Examples of loss prevention techniques*

Loss prevention objective	Technique/approach	
	Manage existing contamination	**Prevent new contamination releases**
Reduce the hazard potential	Convert contamination into a more innocuous substance	Use less hazardous substances in operations Introduce handling protocols for use of hazardous materials
Alter the environment in which the hazard exists	Encapsulate contamination	Bring underground storage tanks and pipework to surface
Alter the processes where hazard and environment interact	Restrict entry to site	Double seal storage tanks holding hazardous substances Construct bunds around storage tanks

3.3.3 Loss mitigation

Loss mitigation measures aim to reduce the severity of a loss by controlling an adverse event as it occurs, as well as the immediate and longer-term consequences of the event. Examples of loss mitigation techniques applied to contamination risks are presented in Table 3.2.

Table 3.2 *Examples of loss mitigation techniques*

Loss mitigation objective	Technique
Limit liabilities created by a new contaminant release	Install, maintain and monitor spill detection instruments Prepare and practice spill containment plan Locate operations in areas of low environmental sensitivity
Minimise sanctions for environmental offences	Show evidence of environmental investments, past compliance and prompt clean up action Demonstrate that the consequences of an incident are not environmentally serious, if possible Seek hardship relief by highlighting any limitations on the company's ability to pay[1]
Minimise Part IIA liabilities	As above, plus identify "appropriate persons" in order to avoid or dilute liability using exclusion and apportionment criteria respectively
Limit the damage to company reputation following an adverse incident	Use public relations consultants to co-ordinate communications between the company and the general public Establish good local relations from the onset
Limit liability under contract and tort	Depending on negotiating position, insert clauses in contract to cap liability and limit the number of parties to whom the company may be held liable
Limit the impact of a default on a loan	Secure the loan, preferably with a fixed charge
Control or reduce impact of legal action	Seek arbitration, mediation or other settlement strategies.

Hardship relief only applies to Part IIA 1990 liabilities, and is not formally specified in WRA 1991 or at common law. It may, however, still be taken into account outside Part IIA EPA 1990.

Organisations that operate diversified businesses are potentially less sensitive to major contamination-related losses than those companies that focus their business exclusively on activities which relate to, or generate risks of, land contamination. This is because diversified businesses may be able to mitigate the financial consequences of contamination at individual sites more effectively by spreading the losses across the business.

3.3.4 Controlled risk transfer

Controlled risk transfer involves the reassignment between organisations of activities that may generate financial losses. Transfer is usually undertaken by subcontracting or outsourcing. For example, a developer (transferor) concerned about potential increased project costs associated with contamination can pass this risk to a subcontractor (transferee) through a fixed price contract. Unless the transfer is not recognised by the courts, it offers complete protection for the transferor. However, this level of protection typically entails an added cost to reflect the costs of the risks being transferred.

3.4 RISK FINANCING

Despite the best risk control efforts, losses do occur. Risk financing methods are used to pay for these losses, as well as funding risk control measures to reduce uncertainty. Financing methods fall into three categories: retention, transfer and composite structures (hybrids), which can be further subdivided as illustrated in Figure 3.4. Details of specific techniques and approaches are presented in Section C.

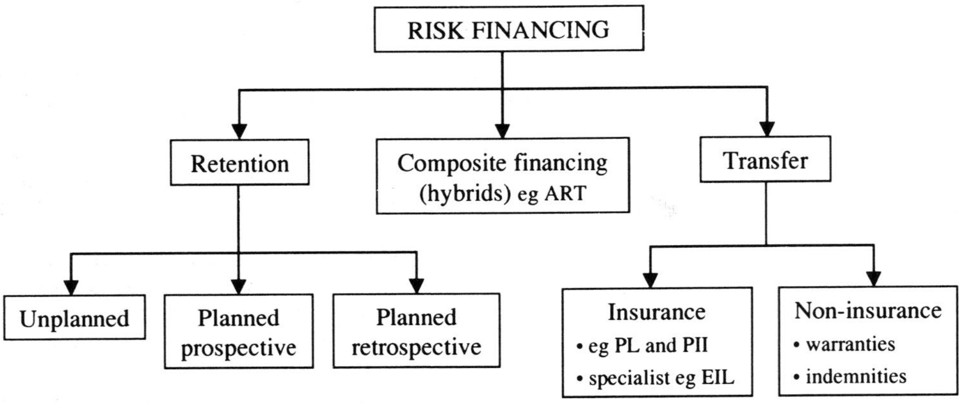

Figure 3.4 *Categories of risk financing techniques*

3.4.1 Retained risk financing

Retained risk financing involves acceptance of risks and financial losses through provision of specific **internal** funds. These funds may be held and administered within an organisation or externally by a third party such as a solicitor. Organisations may plan retention programmes or in the case of unknown risk exposures, retain the risks passively (unplanned). Since the financial risk of land contamination is generally poorly understood, it is anticipated that companies are passively retaining much of this risk.

Risk financing techniques can be classified according to the time periods from which resources are drawn down to pay for losses. Where an organisation has sufficient operating revenue, losses may be absorbed as a current expense. Alternatively, for losses that are large relative to an organisation's financial capacity, there are finite risk instruments to spread the costs over more than one budgeted period.

These methods comprise:

- prospective risk financing where funds are accumulated in advance of losses, ie the funds reflect anticipated losses
- retrospective risk financing where loss payments are spread over time periods following a loss, ie payments reflect past losses.

Formal funding arrangements range from making provisions on the profit and loss account to more sophisticated techniques such as captive insurance. These and other methods are summarised in Table 3.3.

Table 3.3 *Examples of retained risk financing techniques*

Method	Comments
No advance funding (cash flow)	Organisation bears the losses by charging them to current operating costs Cuts administrative costs to a minimum Not suitable for volatile risks with fluctuating losses (such as contamination risks), since the company may experience stress when large losses occur
Liability accounts	Methods whereby an organisation sets up a liability account to absorb fluctuations in uncovered losses. Provisions are added to the account on an annual basis and losses are deducted from the account rather than the organisation's profit Acts to smooth effects on uninsured losses
Asset accounts	Assets or investments, held as cash or near-cash, which can be easily turned into cash to pay for losses Suitable for losses which are likely to exceed internal cash reserves and the amount an organisation could borrow Typically offer a low rate of return
Escrow accounts	Similar to liability and asset accounts, although the account is held with a third party to the risk holder and other organisations which may be affected by the risk
Bonds	Financial instrument used to raise a capital sum which permits the borrower to pay for capital works (either a loss or protective works) and repay the finance over a period usually matched to the period by which the benefit of the works are enjoyed
Finite risk instrument	Structured self-financing plans (banking arrangements) involving the creation of a managed account of money which seeks to stabilise costs of risk over time. The risk holder capitalises the account through premium payments which form the principal source of cash for financing losses. Risk is spread across time rather than exposures Finite risk instrument may be as basic as a line of credit The party managing the account may offer insurance Protects earnings by smoothing major losses over a period of years Funds are invested and a substantial proportion is returned if losses do not occur Any positive fund at the end of the programme can be used to extend the programme, offset future premium costs, or used to fund risk control approaches May be used to manage "uninsurable risks" where the liability is certain but the timing is uncertain
Captives	Allows close control of the precise level of risk which a company wishes to retain Avoids the conventional insurance pricing cycle May provide reduced cost of cover Level of confidentiality higher than conventional insurance Can provide investment income and coverage stability

3.4.2 Financial risk transfer

Financial risk transfer involves the accumulation and use of external funds to the organisation bearing the risk to pay for losses. This is achieved by passing the responsibility to pay for losses from the risk holder (transferor) to a third party (transferee) such as an insurer. This form of transfer differs from controlled risk transfer (Section 3.3.4) in the fact that the risk is not transferred. This has two significant implications for the transferor:

- the immediate consequences of the liability are transferred, but some of the economic burden of the risk may persist eg stigma
- in the event that the transferee defaults on payment of the loss, the liability reverts to the transferor.

Insurance transfers

Insurance is a financial risk transfer mechanism based on a contractual agreement made between two parties: the insurer and the insured (policyholder). Under the agreement, the insurer agrees to indemnify the insured for any losses (as defined in the insurance contract) which the insured suffers. In return for accepting these financial liabilities, the insurer charges the insured a premium.

Insurance has several advantages and disadvantages as a method of risk transfer, as outlined in Table 3.4.

Table 3.4 *Advantages and disadvantages of insurance*

Advantages	Disadvantages
Transfers the risk of unexpectedly large or frequent losses from the insured to the insurer	Some risks – eg criminal fines and existing or imminent losses – are uninsurable. In the latter cases, insurance may be used in an alternative risk transfer (ART) programme to insure the time element component of the risk or to cap the cost of the loss
Substitutes a known cost (premium) for an unknown and highly variable one (losses) Companies can offset the premium as a charge against corporation tax	Not all premiums are certain in timing and amount, and may vary at renewal depending upon the results of the method of calculation eg technical reassessment of a site
Increases certainty in the insured's risk exposure. This may help to facilitate transactions where two parties hold very different views on the risk	Insurance rarely provides full compensation for losses, eg relatively intangible costs, such as stigma and damage to company reputation, will not be covered

Details of the range of insurance products that are relevant to contamination risks are included in Chapter 11.

Non-insurance transfers

Risk allocation is a vital component of many transactions and commissions. Indemnities and warranties serve to apportion risk by transferring liability from one party to another. Since they form part of a contract, warranties and indemnities impose a legal liability on the person giving them. (For definitions see Box 3.2.)

Box 3.2 *Definitions of warranty and indemnity (based on Phillips and Ferera, 1998)*

> A **warranty** is an express or implied statement forming part of a contract that a particular state of affairs exists. It is normally qualified by a letter of disclosure which describes why, in certain cases, the warranty is not true. A person relying on a warranty will not be able to claim breach of warranty for matters disclosed to it.
>
> An **indemnity** is a contractual promise by one party to make good the loss suffered by another party or to pay a defined sum to that other party.

It is easier to claim under an indemnity than under a warranty because a person making a claim under warranty will need to establish the damage or loss suffered before it becomes payable. Under an indemnity, the claimant only needs to show that the event that caused the loss has occurred for the amount indemnified to become payable.

The common law imposes a duty on a party who has suffered loss or damage as a result of a breach of contractual warranty to mitigate its loss. In contrast, no such duty is imposed on a party with a right to claim for loss of damage under an indemnity, unless the contract contains provisions to the contrary (Phillips and Ferera, 1998).

For these reasons, indemnities provide the best means of allocating risk between parties, and are particularly useful if it has not been possible to evaluate the precise extent of the risk (Waite and Jewell, 1997).

Warranties and indemnities have several limitations, including:

- long term protection cannot be guaranteed
  - warranties and indemnities often include time limits and/or sliding scales (where risk is reallocated with time) which may end before the adverse effects of contamination are discovered
  - warranties and indemnities may be secure at the point of transaction, but the creditworthiness of the company providing them may deteriorate to an extent that the company is unable to honour its contractual obligation
- qualification, limitations and caps on liability reduce the benefit of financial protection afforded by warranties and indemnities
- depending on the basis on which damages for breach of warranties are assessed, only limited costs may be recovered. For example, in the case of a share transaction where the vendor provides a warranty that the site is uncontaminated (which is subsequently found to be untrue) the purchaser may suffer significant loss in terms of reduced asset value (including the land). However, if the sale price was assessed according to company profits, the damages that are recoverable under warranty may be negligible, as the profitability of the company may not be affected by the condition of the land
- warranties and indemnities claiming to cover criminal liability are not enforceable in UK courts
- drafting and negotiation can be problematic and time-consuming
- the market will rarely bear the ideal indemnity requirements sought by a purchaser or vendor, or only if counter indemnities are conceded.

3.4.3 Composite financing (hybrids)

In practice, most risk financing programmes are a blend of financial risk transfer and retained risk financing. Organisations may construct their own composite structures, or alternatively seek the services of specialist companies, including insurers, which offer a full range of bespoke financing solutions (see Chapter 10).

3.5 RISK MANAGEMENT IN PRACTICE

With the exception of risk avoidance, it is not possible to eliminate the financial risks posed by contamination. It is inevitable, therefore, that a degree of risk will be borne by one or more companies in the course of land ownership, transaction and development.

3.5.1 Apportionment of risk

Apportionment of risks may be relatively well defined, as in the case of contractual arrangements such as warranties and indemnities. Alternatively the allocation process may be less transparent, with the result that the distribution of risks is unclear and organisations may unknowingly be exposed to risk. The ability to use various risk allocating techniques, such as insurance, warranties and indemnities, will be determined by what the market will bear at that time. Allocation of risk depends on the strengths of bargaining position of each of the parties involved, their willingness to bear risk and tolerate uncertainty, and their capacity to control the risk.

Strength of bargaining

Bargaining strength is influenced by awareness and knowledge of risk. At one end of the scale, organisations can unwittingly bear risk through ignorance. In such cases, it is less likely that they will be protected against any corresponding losses that occur.

Market (supply and demand) circumstances also influence the relative bargaining positions of organisations. Strong demand for land can counteract the effects of contamination and improve the negotiating position of the vendor. The location of a site can, therefore, have a marked effect on the bargaining position of a company.

Certain large, previously polluting industries have been in a sufficiently strong negotiating position, by virtue of a strong demand for their sites and their knowledge of the contamination risks, to require blanket indemnities from organisations purchasing their land.

Willingness to bear risk

A company's willingness to bear risk is influenced by the financial benefit which it receives relative to the degree of risk to which it perceives it is exposed. In principle, a company will be more willing to accept a risk where the return is relatively high, particularly where the risks are readily controlled. Market circumstances may, however, force organisations to accept risks which are not allocated equitably, and where the risks are difficult to control. An illustration of the process of allocating risk for a hypothetical land transaction is provided in Figure 3.5.

The distribution of risks between companies is a dynamic process with risk exposures changing continuously with time as a result of numerous factors including new organisations taking an interest in a property, changes in legislation and operating practices, and changes in the financial standing of organisations.

3.5.2 Current market practices

Based on the survey of the UK market (see Section 1.1), it is apparent that, apart from a few high profile cases, there has been little experience, or at least reported experience, of *material* loss caused by land contamination. Those companies that reported losses claimed that the impact was relatively insignificant in relation to their overall loss exposure – although a few organisations admitted that the cumulative effects of contamination had not been assessed. The apparent lack of financial losses caused by contamination may reflect several conditions at the time of writing, notably:

- the buoyant property market, which acts to reduce the importance of contamination risks (although this effect is subject to considerable regional and local variation)
- the immaturity of relevant UK legislation
- deficiencies in enforcement of environmental legislation.

The financial position of organisations thus may change for the worse when the demand for property declines, legislation matures and/or regulatory enforcement tightens.

The potential impact of new legislation, particularly Part IIA EPA 90, has been the subject of much speculation, with those organisations with a vested interest in contaminated land sometimes guilty of exaggerating the possible effects. The reality of delays in implementing legislation, combined with an absence of high-profile losses has lulled organisations in to a false sense of security. Nevertheless, there is considerable anxiety in the market as a result of the uncertainty.

The approach to financial management of contamination risk has focused, in the first instance, on technical solutions (including due diligence, site investigations, remediation and environmental management) to reduce uncertainty and control the risk. Secondary measures have seen companies relying on traditional insurance policies or, increasingly, warranties (qualified by letters of disclosure) and indemnities to apportion risk. A number of organisations have undertaken controlled risk transfer by outsourcing work that may generate liabilities, although avoidance of contamination risks has rarely featured as the sole motivation behind this move.

Industry has been generally reluctant to go beyond these primary and secondary measures to manage the residual risks financially, or consider alternative methods to finance risks ahead of their crystallisation (Lockett, 1996). It is unclear whether this is due to a market perception that these techniques are unnecessary or flawed in some way, or simply that organisations are not aware of their existence and potential application.

The risk management stance of organisations appears to be changing, as shown by increased activity in the specialist environmental insurance market. Another management option that has engendered increased interest consists of planned and structured risk financing programmes using internal funds and resources. In some cases, insurance has been incorporated within the programme. Whilst these techniques have been used routinely to manage a broad range of non-contamination risks, financing has been used rarely in the context of contaminated land. Notable exceptions include certain major industry sectors, such as oil and chemical, which have operated captives, or the waste management sector which has used "site pensions" to finance the potentially long term risks and liabilities of landfill sites.

Because of the lack of experience of material financial losses, the effectiveness of management techniques has yet to be tested, particularly as a means of long-term financial protection. However, the potential short-term effects of certain techniques, especially insurance, are already becoming evident through the facilitation of land transactions and the realisation of value on brownfield projects.

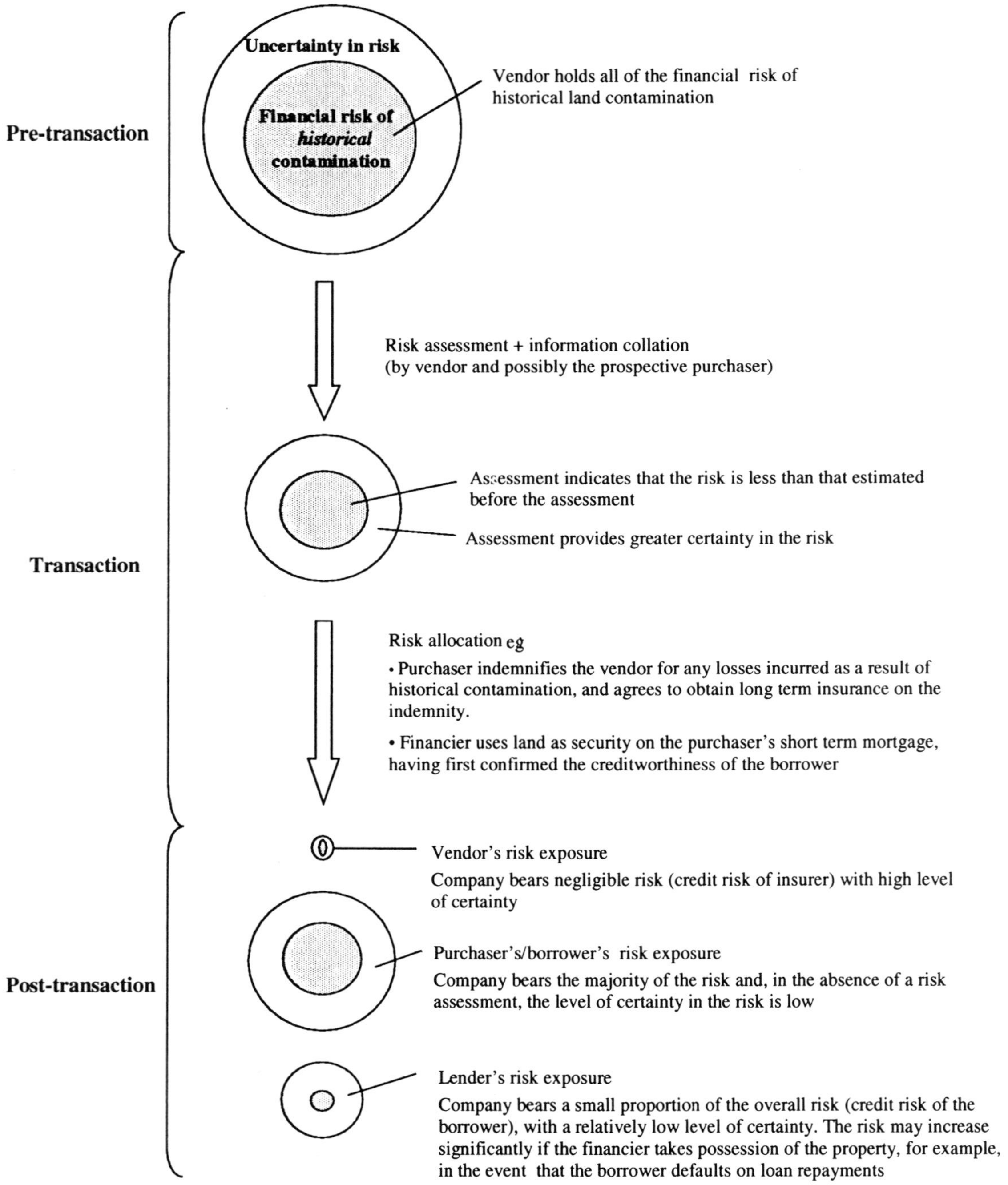

NB: The size of the shaded and clear circle represents the magnitude of the financial risk and the degree of uncertainty in the risk respectively – the greater the size of the circles, the greater the risk and uncertainty.

Figure 3.5 *Illustration of the process of risk apportionment for a hypothetical land transaction*

SECTION B – GUIDANCE

4 Overview of guidance

4.1 FRAMEWORK

The following five chapters provide guidance on how to manage the financial risks from land contamination according to the activities by which organisations are exposed to risk.

Chapter 5	Land ownership and occupancy
Chapter 6	Land transactions (mergers, purchases, sales and leases)
Chapter 7	Development and contracting
Chapter 8	Financing and investment
Chapter 9	Provision of, and reliance on, professional advice

Each chapter encompasses a review of key legal issues, a table identifying the specific risks and a description of management options. Generic aspects of the guidance that are common to all activities, such as cost benefit analysis and development of a management strategy, are described in this chapter. Section C provides more detailed information on financial management techniques.

4.2 METHODOLOGY

The approach to financial risk management is based on a multi-stage process. The key steps are summarised in the flow chart in Figure 4.1 and described below.

4.2.1 Step 1 – Screen business risks (optional)

Before assessing contamination risks in any detail, the first step comprises an optional screening exercise to consider the impact that contamination may have on a business in terms of:

- increasing expenditure
- reducing income
- impairing asset value
- increasing liabilities.

The screening process provides an indication of those areas of the business which are more susceptible to financial losses arising from land contamination, thereby allowing the organisation to a) focus its resources during the subsequent risk assessment and management steps on those areas of the business which warrant greatest protection; and b) approach the risk management process from a strategic business-objectives perspective.

The screening process may be undertaken in the absence of any supporting technical data on contamination, and can be applied to specific projects or entire business entities or streams. Details on the use of similar approaches to financial risk evaluation are provided in *Risk analysis and management for projects* (Institution of Civil Engineers and the Faculty and Institute of Actuaries, 1998).

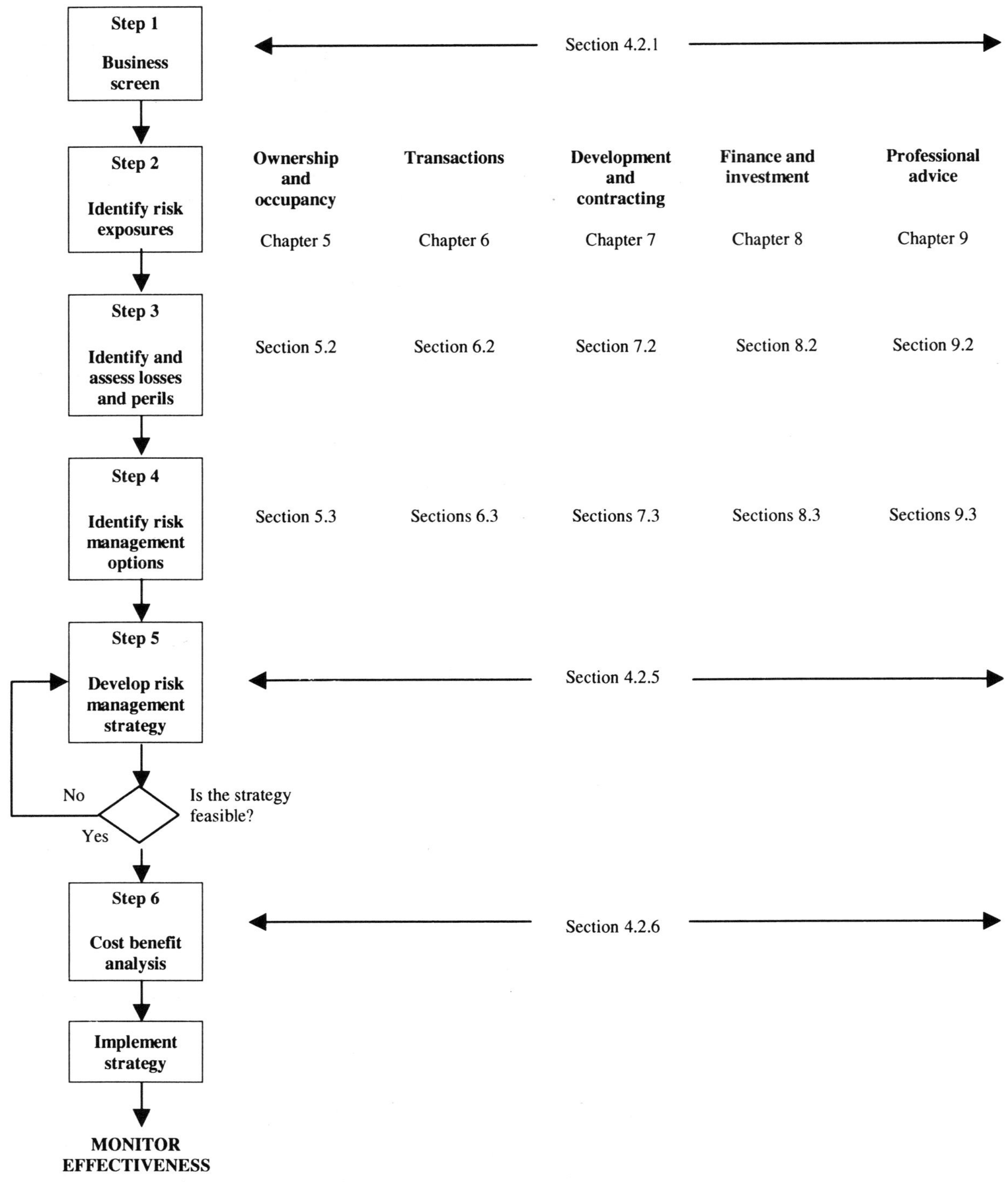

Figure 4.1 *Framework for managing the financial risks of land contamination*

An example of the application of the screening process to the following three hypothetical businesses is described in Table 4.1.

1. A **contractor** involved in land remediation projects. The turnover is assumed at £11.5 million per annum, with costs of £10.5 million per annum. Balance sheet of the company is approximately £1million.
2. A **financier** involved in providing debt finance to companies that might be exposed to contaminated land risks. The loan book to clients with risks in the land contamination sector is assumed at £100 million.
3. A **developer** of commercial projects on contaminated land sites. The developer has £20 million assets with an annual turnover of £5 million.

Table 4.1 *Illustrative example of business screening process*

Organisation	Nature of financial risk	Relevant figure (£ 000)	Reasonably foreseeable loss (% of relevant figure)	Probability of loss (%)	Risk to business (£ 000)	Ranking
Contractor	Increased expenditure	£10 500 (cost of sales)	5	15	79	1st
	Loss of income	£11 500 (sales)	<1	<5	6	2nd
	Decrease in balance sheet value	£1000 (balance sheet assets)	<1	<1	~0	=3rd
	Increased liabilities	£1000 (balance sheet liabilities)	<1	<1	~0	=3rd
Financier	Loss of principal	£100 000 (loans outstanding)	2	5	100	1st
	Loss in income	£6000 (interest payments and fees)	1	1	1	3rd
	Increased liabilities	£100 000 (liabilities)	2	3	60	2nd
Developer	Loss of income	£5000 (turnover)	20	20	200	2nd
	Increased expenditure	£3000 (investment)	20	30	180	3rd
	Decrease in balance sheet value	£20 000 (balance sheet assets)	20	10	400	1st
	Increased liabilities	£20 000 (balance sheet liabilities)	10	5	100	4th

The analysis highlights three points.

1. The contractor should be most concerned with ensuring that expenditure does not increase without an equal increase in the relevant levels of income generated. Securing control over contaminant-related expenditure might be achieved most effectively by insurance or contractual arrangements.
2. The main area of risk for the financier is loss in principal. The financier might, therefore, seek to secure its position by undertaking comprehensive due diligence before entering into a project in order to assess the viability of the project on which a loan is being provided. The financier might also restrict the actions that the lender can take in respect of increasing the potential exposure to risks.
3. The main risk management objective of the developer is limiting the reduction in balance sheet value. This may be achieved in the same way that the developer secures its asset values. For example, securitisation of income flows from properties away from the underlying assets might help to reduce the potential exposures to contaminated land risks.

As a next step in undertaking further investigations, investment should be focused towards risk protection in those areas which help to reduce risk in the areas of greatest cost of exposure to the business. In all respects, however, the options which deliver the best value for money should be the priority areas for investment.

4.2.2 Step 2 – Identify risk exposures

Decide which activity is relevant to your situation and turn to the appropriate pages in Chapters 5–9 (see Figure 4.1).

4.2.3 Step 3 – Identify and assess losses and perils

From the relevant sections in Chapters 5–9 (see Figure 4.1).

4.2.4 Step 4 – Identify risk management options

From the relevant sections in Chapters 5–9 (see Figure 4.1).

4.2.5 Step 5 – Develop risk management strategy

Strategy objectives

Corporate risk management strategies should aim to manage existing contamination liabilities whilst avoiding creating new ones.

This can be achieved through:

- optimising environmental performance – creation of future liabilities through existing practices may be kept to a minimum by optimising environmental performance and maintaining good housekeeping
- adopting a cautionary approach to new business activities – exposure to new risks can be minimised by appropriate structuring and vetting of transactions and other new activities
- developing a strategy to manage residual risks and contingent liabilities.

Strategy options

There are four general strategies to financial risk management:

- do nothing
- focus on the risk
- focus on the liability
- focus on the risk and liability.

Strategy 1: Do nothing

This is a valid strategy under a limited set of circumstances, for example, where the company operates a diverse business portfolio, has significant contingent reserves set aside for catastrophes and contamination presents a minor contribution to the aggregate business risk. Diversion of reserves from other more important areas of the business to fund contamination risk management may cause damage to the company.

Strategy 2: Focus on the risk

This is a proactive, preventative management strategy which is particularly applicable to controllable risks which may readily escalate. The strategy encompasses loss prevention techniques to reduce the probability that a loss may occur. The residual risk is retained passively, ie plans are not necessarily made for financing the losses, as shown in Figure 4.2.

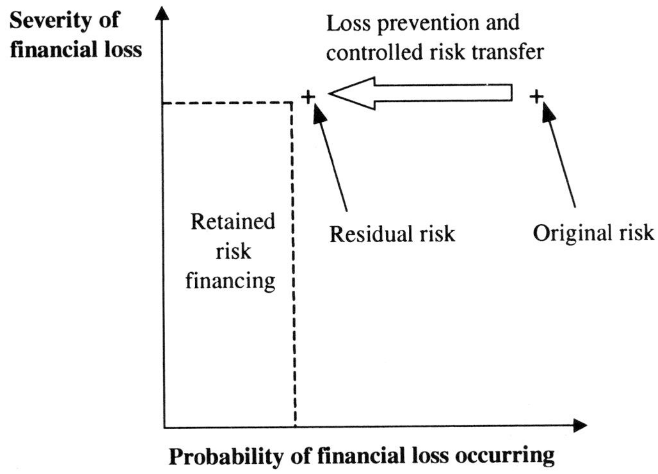

Figure 4.2 *The effect of a "focus on the risk" approach on financial risk*

Strategy 3: Focus on the liability

This is a reactive management strategy which accepts that losses will occur. The strategy encompasses loss mitigation mechanisms and/or risk financing (see Figure 4.3). A limited risk assessment may suffice, so long as the company is in a strong position to finance the losses when they occur. Otherwise, it is prudent to undertake more detailed risk assessment to provide more confidence that there is adequate funds to cover losses if and when they occur. This strategy is applicable to risks which are difficult to control, for example managing contingent liabilities of land which was formerly owned by the company.

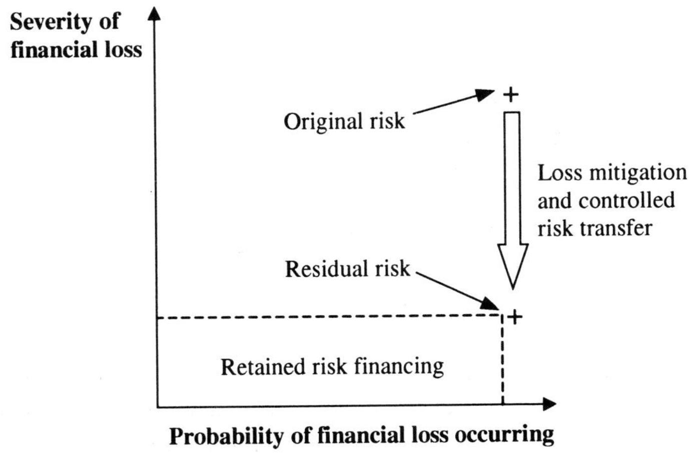

Figure 4.3 *The effect of a "focus on the liability" approach on financial risk*

Strategy 4: Focus on the risk and liability

Apart from avoiding the risk in the first place, this strategy offers maximum protection against financial losses, but at the greatest cost. It is particularly appropriate for uncertain risks which contribute significantly to a company's aggregate business risk. The approach commonly utilises a combination of technical, legal and financial measures to control and finance the risk, backed up by comprehensive risk assessment to provide greater confidence in the appropriateness of the management strategy (see Figure 4.4).

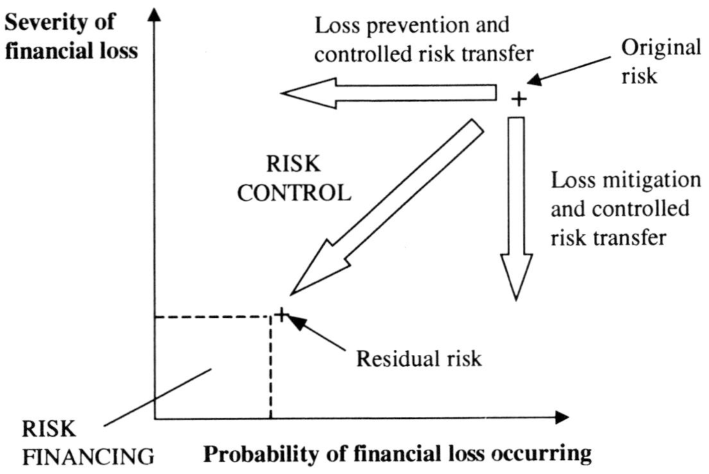

Figure 4.4 *The effect of a focus on the risk and liability approach on financial risk*

Strategy development

With only limited resources normally available to manage business risks, there needs to be prioritisation in any risk management programme. The factors outlined in Table 4.2 will collectively influence the level of expenditure, if any, that a company chooses to spend on contamination risk management, and the type of strategy it may develop.

The feasibility of a risk management strategy depends upon the availability of suitable techniques at an affordable price. By analysing the factors shown in Table 4.2, certain risk management approaches and techniques can be eliminated and the feasibility of the overall strategy determined. Retention may be the only feasible approach in situations where the organisation cannot control the loss, avoidance is impossible and no transfer options exist.

Development of a working strategy is an iterative process, since the strategy may not be feasible, or may only be possible at an unacceptably high cost. The feasibility may also change over time.

Table 4.2 *Suggested criteria for use in developing a risk management strategy*

Development criteria	Examples
Is contamination risk an ***important*** part of aggregate business risk?	Companies with poorly diversified business activities which are potentially affected by contamination are particularly sensitive to the risk. These companies should, as a priority, invest in risk management measures that address the risk, both in terms of cumulative and catastrophic losses. This could be achieved, eg, by purchasing insurance with aggregate deductibles. Other businesses that show greater diversity in their business activities are able to spread contamination-related losses across the business portfolio. The risk may be less important to these companies, but should still, nevertheless be managed. A company's perception of the importance of contamination risk will be influenced by a number of factors including its awareness of the risk and its experience of losses (as a company or an industry sector). Contamination risk is often high impact, low probability, so an absence of reported losses might provide a false sense of security. External factors, including changes in the property market, will also have an effect on a company's perception of the importance of contamination.
What is the company's ***business motivation***?	A company's business motivation will also influence the strategy develop-ment. Risk retention may be the preferred option where the motivation is short-term gain. Where long term survival is the objective, a degree of risk transfer will be preferable.
Is the risk ***avoidable***?	Risks associated with core business activities, particularly the more profitable ones, are less conducive to a risk avoidance strategy than fringe activities offering lower returns.
Can the risk be ***controlled***?	Contingent liabilities associated with the former ownership of land are largely uncontrollable thus promoting a financing-based strategy.
Can the ***risk activity*** be ***transferred***?	Controlled risk transfer is usually only appropriate when the party taking on the risk is in a strong position to control it (eg specialist subcontractor). The costs of transferring the risk in other cases may be prohibitively high.
Can the risk be ***transferred financially***?	Certain risks cannot be insured, for example: fines; losses which are already occurring or which are imminent; and intangible losses such as blight and damage to reputation. The ability to transfer risk by non-insurance methods (contract) will depend upon the relative bargaining positions of organisations.
Can the risk be ***financed internally***?	Sufficient capacity for bearing loss is an essential requirement for retained risk financing. This capacity must be sufficient to bear the maximum probable cost associated with the risk, or retention becomes unwise. Loss bearing capacity does not necessarily need to be viewed in the context of a single budget period.

4.2.6 Step 6 – Cost-benefit analysis

Having developed a risk management strategy, the cost-benefit analysis step aims to establish the most effective way to implement the strategy. The analysis can be undertaken very simply by scoring each risk management option against the following criteria, using weighting factors where necessary to reflect the relative importance of each of the following criteria to the company:

- costs and benefits
- durability
- tax implications
- added value.

The effects of each of these criteria on the selection process are outlined below.

Cost and benefits

The benefits of financial risk management can be viewed in terms of the potential loss avoidance for a fixed cost. For risk control techniques (such as remediation), the cost/ benefit ratios can be significant, although the curve tends to flatten out quickly (law of diminishing returns). The intangible benefits of risk management, such as improved stakeholder perception, should not be overlooked despite the difficulties in quantifying the effects in financial terms. Companies should not only look at the costs of undertaking various risk management techniques, but also the costs of not implementing them. The time and duration of the expenditure should also be considered ie the time value of money. Companies may also consider the returns that are generated from investing in the management of contamination risks, compared to the returns that would be generated from investment in other areas of the business.

Durability

The effectiveness of risk management techniques will change with time. This may be very obvious as in the case of an indemnity written on a sliding scale, or more subtle as in the case of a remediation project designed to current standards (which may prove to be inadequate in future years). It is important therefore that the timescale of the protection is taken into account when selecting appropriate techniques.

Tax implications

The specific tax treatment of any expenditure that a company might make on risk management will depend on many factors, including the profitability of the company, its business and the nature of the adopted risk management technique.

Added value

Certain techniques offer specific benefits to a company other than the management of risk and uncertainty. Such added value elements may include:

- off-balance sheet management of liabilities – this is particularly attractive where a company may wish to hide the damaging effects of a liability
- attraction of investment in contaminated sites, facilitation of deals and avoidance of "deal breakers" and enhanced realised value of land – the increased certainty that insurance introduces can effectively bridge gaps in risk perception between parties
- the benefits of insurance can be assigned to third parties under certain circumstances.

5 Land ownership and occupancy

This chapter identifies the contamination-related financial risks of land ownership, occupancy and associated activities, together with options to manage these risks. Chapter 4 (Section 4.2.5) presents an approach for using this information to develop an appropriate risk management strategy.

5.1 BACKGROUND

An organisation's interest in land may include any of the following:

- ownership and occupancy of a freehold (owner)
- ownership of part or all of a freehold, and leasing of the property to a third party (landlord)
- occupancy of a site under a short term lease (occupier/tenant) or a long term lease of greater than 21 years (owner)
- occupancy of all or part of a site as a subtenant
- management of a site as a licensee
- control of a property as a trustee
- control of a property as an administrator or receiver.

This chapter addresses the management of contamination risks associated with **current** and **historical** interests in a site. Chapter 6 provides guidance on how to minimise the retention or acquisition of contamination risks during **future** transactions of assets and shares. Chapter 7 provides guidance on how to manage the financial risks of redevelopment and remediation of contaminated sites.

5.1.1 Stakeholders

The key stakeholders in land ownership and occupancy include the following:

- risk holders: current landowners, including homeowners and investors; tenants and subtenants; licensees; trustees and executors (ie parties with direct control over the land)
- third parties who may potentially hold or acquire primary liabilities but who would not normally exercise direct control of the risk: former owners and occupiers; and financiers (see Chapter 8). Freeholders of land which is subject to a lease may also fall under this category
- other parties: professional advisers; regulators and the general public.

5.2 RISK EXPOSURES

The financial risks that contamination presents to landowners, landlords, tenants, subtenants, licensees, trustees and receivers are summarised in Table 5.1 and discussed further in Sections 5.2.1–5.2.7.

Table 5.1 *Financial risk exposure of organisations with a controlling interest in land*

Financial impact (if risk materialises) / **Financial risks presented by contamination**	**Unforeseen additional operating costs**[1]	**Asset impairment**[2]	**Enforced remediation**[3]	**Third party liabilities**	**Contractual liability**	**Personal liability**	**Legal costs**	**Fines, injunctions and prosecutions**	**Reduced revenue**[4]	**Damage to corporate or brand reputation**	**Key factors affecting the risk**
Landowner (and occupier)											
Increased liabilities	✓	✓	✓	✓	✓	✓	✓	✓	✓	✓	Uncertainties surrounding the financial impact of: – possible historic contamination at the site – new pollution from current operations – potential for contamination to migrate on to site from neighbouring premises – residual contamination at sites formerly owned/occupied by the company
Adverse performance of business operations	✓	✓							✓	✓	As above
Landlord											
Increased liabilities		✓	✓	✓		✓	✓	✓	✓	✓	Potential of tenant's activities to cause contamination Uncertainties of effectiveness of leases in protecting landlord against liabilities arising from tenant's activities
Adverse impact on value of freehold in reversion	✓	✓							✓		As above
Tenant/subtenant											
Increased liabilities			✓	✓	✓	✓	✓	✓		✓	Uncertainties surrounding the financial impact of possible historic contamination at the site, and the effectiveness of lease terms in protecting the tenant from acquiring such liabilities
Adverse impact on tenancy	✓								✓		As above
Licensee											
Increased liabilities			✓	✓	✓	✓	✓	✓		✓	
Trustee											
Increased liabilities			✓	✓	✓		✓	✓		✓	Uncertainties surrounding the trustee being construed as a polluter or innocent owner
Administrator/receiver											
Acquisition of liabilities			✓	✓	✓		✓	✓		✓	Uncertainties surrounding the receiver's actions which could be construed as causing or knowing permitting contamination

Notes

1. Eg costs to prevent liabilities, for example, by upgrading plant, or additional costs of securing finance and insurance.
2. Eg loss of land value, inability to use the asset.
3. Under Part IIA EPA 1990, WRA 1991 or common law.
4. Eg loss of shareholder value, loss of market.

5.2.1 Liabilities

Under certain circumstances, landowners (including homeowners), occupiers, tenants, subtenants, licensees, trustees and receivers can all, by virtue of their ownership or occupancy interests and/or their potential to cause or knowingly permit contamination at a site, incur the following liabilities (see also Section 2.4).

1. **Civil liability**
   Third party liability may be incurred for damage caused, or likely to be caused, by contamination. In such cases, a third party may seek compensation for its loss and require the polluter or owner/occupier to take steps to prevent future contamination.
2. **Remediation liability**
   An organisation may incur remediation liabilities in the following capacities:
   a) as a polluter, causing or knowingly permitting the *entry* of pollutants into water (WRA 1991) or the disposal of waste (Part II EPA 1990); or causing or knowingly permitting the **presence** of substances which cause, or threaten, to cause significant harm (Part IIA, EPA 1990). In the latter case, liability extends not only to those companies that knowingly permitted the presence of substances from the outset, but also those that knowingly permitted their continued presence. Among persons who have caused or knowingly permitted the presence (or continued presence) of substances, the guidance under Part IIA channels liability towards those who gave rise to changes in substances, introduced pathways and/or introduced receptors. Thus, any company will fall into the "polluter" liability group if they have knowledge of and control over substances which are linked to a significant pollutant linkage. The associated liabilities may remain with the company, even if the problem manifests itself after they have ceased to have an interest in the site.
   b) as a current owner/occupier under Part IIA EPA 1990 where no person who has caused or knowingly permitted can be found (but not for liability for pollution of controlled waters or certain cases where the contaminants have migrated).
3. **Contractual liability**
   Contractual liability can be incurred through provisions and conditions in contracts, such as covenants, warranties and indemnities.
4. **Criminal liability**
   Polluters can be fined and prosecuted for environmental offences, such as pollution of controlled waters. Personal liability may extend to directors, officers, partners and, under certain circumstances, those with a non-executive role.

5.2.2 Landowner's financial risk exposure

A landowner's contamination-related financial risk exposure comprises liabilities and disruption to business operations.

Liabilities

A landowner can be held legally liable for:

- new contamination from current operations at the site
- historic contamination which pre dates the company's interest in a site
- off-site migration of new or historic contamination on to neighbouring premises
- on-site migration of contamination from neighbouring premises
- contamination on land owned or occupied previously by the company.

Liabilities will have an adverse financial impact on a company through direct losses, such as fines, damages or remediation costs. Other losses may arise indirectly as a consequence of being held legally liable, for example:

- disruption of business operations during remediation
- diversion of management time and resources away from profitable areas of the business
- adverse effects on share price
- stigma leading to reduced land value
- increased costs of purchasing insurance and obtaining finance in the future.

Adverse performance of business operations

Even if contamination does not trigger legal liabilities, it may still impact financially on an operating business through the extra costs to manage the contamination and avoid escalation of the risk (for example by monitoring contamination or installing containment systems such as barrier walls). The presence of contamination may limit the development potential of the land, reducing its marketability and potential value. The mere presence of contamination can also still generate adverse reaction amongst stakeholders.

5.2.3 Landlord's financial risk exposure

A landlord's financial risk exposure comprises increased liabilities and adverse effects on the lease. Exposure is influenced by the terms of the lease.

Liabilities

A landlord can acquire liabilities as a result of the unauthorised polluting activities of its tenants or sub-tenants. These activities may be legal (indeed permitted at the time), illegal, legal but contrary to the terms of the lease, known during the lifetime of the lease, identified on termination of a lease or discovered at some later date.

Adverse effects on the lease

A landlord may face additional operating costs as a result of the need for more frequent and thorough due diligence of its tenant's activities, or due to remediation after the lease ends. If the tenant or sub-tenant knowingly or unknowingly contaminates the site and that contamination is discovered at some later date, the value of the land may fall. If contamination disrupts the tenant's business activities, the landlord faces possible delays in payment of rent. Where contamination causes the tenants' business to be adversely affected, or even to fail, the landlord can suffer a loss of rental income.

A landlord's company name may also become associated with a contaminated site, even though the company neither caused contamination nor is paying for its remediation. The association may damage the company's corporate or brand image, which may be reflected as a reduction in share price.

5.2.4 Tenant's financial risk exposure

A tenant's financial risk exposure comprises increased liabilities and adverse effects on the tenancy. Exposure is influenced by the terms of the lease.

Liabilities

A tenant may acquire liabilities for contamination that it did not cause. For leases of more than 21 years, the landlord's liability as a Class A appropriate person under Part IIA EPA 1990 may pass to the tenant under the "sold with information" test (but only if the tenant knowingly permits the continued presence of the contaminants).

Many leases also make the tenant responsible for complying, at its own cost, with all statutory requirements or notices served on anyone in respect of the property. Alternatively, the landlord can recover compliance costs through a service charge. Including a covenant to ensure that the property is no more contaminated at the end of the lease than at the beginning may, depending on the terms of the covenant, also make the tenant liable for any contamination which migrates on to the property from elsewhere during the tenancy. Covenants such as these can effectively make the tenant contractually liable for contamination that it did not cause. If such covenants are included, a baseline survey should be conducted of the state of the property at the beginning of the lease.

A tenant's own actions may cause contamination, leading to a breach of the terms and conditions of the lease and possible criminal, civil and remediation liabilities. Officers of the tenant organisation may also be found personally liable. Legal costs may be incurred defending claims.

Adverse effects on the tenancy

A tenant may face additional unforeseen operating costs as a result of the need for more comprehensive environmental management, or due to remediation during the lease. The corporate reputation of a tenant may also be tarnished if the company is associated with a contaminated site, even if it neither caused the contamination nor is responsible for its remediation. In turn, damage to reputation can have an adverse impact on share price.

5.2.5 Licensee's financial risk exposure

A licensee generally has less financial risk exposure than a tenant due to the licensee's lesser interest in a site. In addition, exclusion tests under Part IIA EPA 1990 make it more likely that a licensee will be excluded from liability than a tenant.

5.2.6 Trustee's financial risk exposure

A trustee's financial risk exposure is strongly influenced by uncertainty concerning the trustee's status as a causer or knowing permitter of contamination. In addition, a trustee is specifically defined as an owner under Part IIA EPA 1990. Recent case law appears to indicate that trustees may ensure that trust assets are retained to cover any costs arising out of Part IIA EPA 1990 related liabilities.

5.2.7 Administrators financial risk exposure

Administrators and other insolvency practitioners such as receivers will not be personally liable under Part IIA EPA 1990 unless the contamination arises from their unreasonable act or omission. They may, however, possibly be liable under other environmental legislation or at common law if, for example, they cause pollution.

5.3 MANAGEMENT OPTIONS

The following sections outline risk management options for landowners, landlords, tenants and trustees.

5.3.1 Landowner's options

The objective of a landowner (including those companies under a long-term lease) is to manage the risks of historic and new contamination at its existing sites, together with the residual risks of contamination at sites formerly under the company's ownership or occupancy. The following techniques can be used singularly or in combination to help meet these objectives:

- implement environmental management systems (loss prevention and loss mitigation)
- proactively investigate, monitor and break pollutant linkages (loss prevention)
- dispose of site (controlled risk transfer)
- develop defence plan (loss mitigation)
- diversify activities and restructure the business (loss mitigation)
- finance losses through provisions, captives, finite risk instruments and other funding mechanisms (retained risk financing)
- finance losses through composite structures (composite risk financing)
- purchase insurance (financial risk transfer).

Environmental management system (EMS)

Environmental management encompasses policies, procedures and operating practices designed to reduce the likelihood of environmental damage occurring and/or to mitigate the impact of damage. For example, construction of bunds around above ground storage tanks can prevent financial losses arising as a result of contamination entering the ground. Similarly, installation of an overspill or a leak detection system on above ground or underground storage tanks can mitigate the damage caused by an accidental release, by triggering a prompt management response. An EMS is normally adopted for the management of risks arising from new contamination, but may equally be applied to historic contamination (see below).

In addition to risk control, environmental management can offer business benefits such as possible cost savings through improved operating efficiencies and reduced frequencies of supplier audits and inspections by the regulators. An EMS can also enhance an organisation's external perception, particularly if the system is accredited to a certified standard such as ISO14001 or EMAS. Certain investors have also recognised the value of tracking environmental performance as an indicator of a well-managed company. The main limitation of environmental management is that systems are not fail-safe and are often affected by human errors.

Investigation and remediation of historic contamination

In the case of a historic contamination, proactive environmental management (site investigation, risk assessment and, where appropriate, remediation) can minimise escalation in remediation costs (see Section 3.1.2), as well as reducing the likelihood of incurring liabilities.

Remediation may take the form of treatment of a contaminant source (eg by excavating and disposing of contaminated soils to landfill or using process-based treatment technologies) or interception of a pathway (eg by installing a containment barrier wall). Following remediation, the property may become more marketable, increasing its value.

The main limitation of this approach is the uncertainty surrounding the costs and effectiveness of investigation and remediation. This uncertainty can be addressed by various "back-up" management techniques, including warranties and remediation cost cap insurance, as discussed in Chapter 7.

Disposal of site

Selling a site will reduce a company's risk exposure, although, depending on how the sale took place, the company may still retain liabilities for contamination present at the point of sale. By relinquishing the site, the company reduces the degree of control it has over the risk. As a result, financing may be the only effective "back-up" option to manage post-transaction liabilities. Section 6.3.1 provides details on minimising the retention of risk during the sale of a site.

Defence plan

The objective of a defence plan is to mitigate any losses arising from contamination. It is an approach that is equally applicable to currently owned sites and those no longer under the control of the company. Information management is an important component. Examples of defence plans include:

- forming a strategy to respond to a regulator's powers of search, seizure and questioning
- seeking legal privilege for sensitive documents to protect them from disclosure in common law
- building up a dossier of evidence which can be used to challenge statutory notices on grounds that the land is not contaminated, the company is not the person who caused the contamination or is excluded from liability under Part IIA, or remediation requirements are unreasonable
- using public relations consultants to control communications to the press and general public on any potentially damaging incident in which the company is implicated.

Corporate restructuring and diversification of business activities

Diversification of the business into activities that are not affected by, or are less sensitive to, land contamination will spread the risk across a portfolio of operations, thereby reducing the relative importance of contamination to the company.

As a note of caution, directors can be held personally liable under the Insolvency Act 1986 and associated legislation, where an organisation transfers its assets into a subsidiary or nominee company with the intention of limiting liability to the value of the relevant property (where the nominee company has no further assets). LAs and others may apply to a court for an order to reconstitute or reinstate a company which has been dissolved.

Retained risk financing and composite structures

Retained risk financing includes building up funds prospectively or retrospectively to pay for losses. The approach is applicable to all types of financial risks and is particularly relevant to contamination risks over which a company may have restricted control, such as those associated with former ownership of land. The main limitation is the uncertainty of setting appropriate financing terms. If the terms are inadequate, the company may suffer unplanned impact on its profit and loss. If the terms are set too generously, funds are drawn unnecessarily from other areas of the business, which may suffer as a result.

Blending financing with risk transfer as part of a composite structure can reduce the uncertainty of setting appropriate financing terms. This approach can provide budget stability for financing losses that are inevitable but the timing and cost are uncertain. The premiums may be tax deductible and, in the event that a loss does not materialise, a large proportion of the fund is returnable. Contamination risk exposures can be one of a number of exposures blended into a composite structure, which improves pricing and coverage efficiency. Retained risk financing and composite structures are described in Chapters 10.

Insurance

Insurance can provide financial protection against most losses arising from historic, new or residual contamination. A review of insurance products is included in Chapter 11.

5.3.2 Landlord's options

Remedies are aimed at minimising the risk of a landlord being held liable for contamination, and preserving the value of the freehold in reversion. There are four management approaches:

- enforcement of existing lease provisions, obligations and covenants
- definition of new provisions and obligations
- instigation of criminal-based action by the regulators
- use of tortious rights.

Enforcement of defined provisions and obligations in leases and tenancies

Old leases may not provide adequate protection against losses arising from environmental contamination caused by a tenant's activities. Therefore, lease conditions should be reviewed to establish the degree of contractual protection that is currently provided. Key conditions and provisions are discussed in Section 6.3.3. In all cases, specialist legal advice is recommended.

Terminate present leases and renewal

In the event that existing leases do not protect the landlord, leases may rarely be terminated and renewed on more favourable grounds to the landlord (if there is consideration for doing so). This is easier to carry out under Scottish law than it is under English law (which provides business tenants with rights of renewal). More realistically, a landlord may renew the lease on more favourable grounds at the end of the lease period (see Section 6.3.3). In considering terminating or forfeiting a lease contract, attention should be paid to obligations incumbent upon the tenant at termination of the lease.

Criminal-based action by the regulator

Under certain circumstances it may be worth considering instigating regulatory action to prevent an ongoing problem. Extra caution should be exercised in alerting regulatory authorities to problems, since doing so may lead in certain cases to the landlord being liable, for example, as an owner under Part IIA EPA 1990 (see Section 2.4).

Civil action

Irrespective of the terms of a lease, a landlord may have a claim for damage to reversion if the land is impaired. This may be important where the lease contains no express or implied provision relating to contamination-related damage.

5.3.3 Tenant's options

Remedies are aimed at managing potential losses arising as a result of a tenant's polluting activities and contractual liabilities for historic contamination. In the former case, management options include the following techniques, as described in Section 5.3.1:

- implement EMS (loss prevention and loss mitigation)
- finance risk through provisions, captives, finite risk instruments and other funding mechanisms (retained risk financing)
- finance risk through composite structures (composite risk financing)
- purchase insurance (financial risk transfer).

Tenants should also review conditions on their existing leases to assess contractual liabilities for contamination which pre dates their occupancy of a site. In the event that existing leases impose unacceptable liabilities, a tenant can seek to terminate the lease and renew it on more favourable grounds, or more realistically, seek to renew the lease on more favourable grounds at the end of the lease period (see Section 6.3.4)

5.3.4 Trustee's management options

In as much as a trustee is the owner of the property, he should consider the management options outlined in Section 5.3.1. In addition, a trustee may wish to consider seeking judicial approval for his actions, as was done in *X* v *A, B and C.*

6 Transactions of assets or shares

This chapter identifies the contamination-related financial risks associated with asset and share transactions, together with options to manage these risks. Chapter 4 (Section 4.2.5) presents an approach for using this information to develop an appropriate risk management strategy.

6.1 BACKGROUND

Inevitably transactions change the financial risk exposures of purchasers and vendors. When the share capital of a company is sold, the purchaser automatically acquires continuing commitments and liability for all past environmental transgressions of the vendor, whether or not they are disclosed in the sale process. The transfer includes any contingent liabilities associated with previous site ownership and occupancies. In contrast, the transfer and allocation of liabilities is less clear-cut for asset transactions.

6.1.1 Stakeholders

The key stakeholders in a transaction include:

- risk holders: vendors and purchasers; landlords (lessors) and tenants/subtenants (lessees)
- potential holders of primary liability: former owners and occupiers of the site; financiers; investors, guarantors and insurers
- other third parties: professional advisers; regulators and the general public.

This chapter focuses on the risks and liabilities of purchasers, vendors, landlords and tenants.

6.2 RISK EXPOSURES

The financial risks that contamination presents to vendors and purchasers are summarised in Table 6.1 and discussed further in Sections 6.2.1 and 6.2.2. The risks for landlords and tenants entering into a new lease are similar to those for vendors and purchasers respectively. The main difference lies in the ongoing legal relationship in the form of express lease provisions, which determines the future conduct and liability of landlord and tenant. These provisions can give rise to contractual liabilities. Management of contractual liabilities is discussed for landlords and tenants in Sections 6.3.3 and 6.3.4 respectively.

6.2.1 The vendor's financial risk exposure

Land contamination presents the following financial risks to a vendor:

- protracted or abortive negotiations
- reduced sale price
- contingent liabilities (post-transaction).

Table 6.1 *Financial risk exposure of purchasers and vendors*

Financial risks / Financial impact (if risk materialises)	Wasted expenditure[1]	Increased transaction costs/reduced profit	Extra operating costs[2]	Site not suitable for intended use	Reduced income flow	Enforced remediation[3]	Third party liability[4]	Contractual liabilities[5]	Fines and prosecutions	Asset impairment[6]	Damage to reputation	Key risk factors
Vendor												
Protracted negotiations and/or aborted transaction	✓	✓										Differing risk perceptions of parties Turnaround of project[7]
Achievement of below-optimum sale price		✓										Differing bargaining strengths of parties to transaction which depends on the relative quality and quantity of information held by vendor and purchaser
Retention of contingent liabilities			✓			✓	✓	✓			✓	Uncertainties in the risk of future impact of, and responsibility for contamination
Purchaser												
Protracted negotiation and/or aborted transaction	✓	✓										Differing risk perceptions of parties
Adverse unforeseen effects of contamination on business operations			✓	✓	✓					✓	✓	Uncertainties in the nature, extent and hazard potential of existing contamination
Acquisition of liabilities (not fully taken into account in purchase price)			✓			✓	✓	✓	✓	✓	✓	Uncertainties in the nature, extent and hazard potential of existing contamination Uncertainties in the effectiveness of mechanisms to allocate risk

Notes

1 Eg costs of undertaking due diligence.

2 Eg costs to make site fit for intended purpose; undertake long term monitoring or preventative remediation (in order to reduce the risks of incurring future liabilities or suffering escalating contamination problem); or possible extra costs associated with securing finance or insurance.

3 Part IIA EPA 1990, WRA 1991 and common law.

4 In respect of bodily injury and property damage.

5 The risks for landlords and tenants entering into a new lease are similar to those for vendors and purchasers respectively. The main difference lies in the ongoing legal relationship in the form of express lease provisions, which determines the future conduct and liability of landlord and tenant. These provisions can give rise to contractual liabilities.

6 Eg reduced development potential of the site.

7 Fast turnaround transactions are likely to be more sensitive to the effects of contamination detected during negotiations.

Protracted or abortive negotiations

Vendors and purchasers often have different perceptions of the risk posed by a site. This can lead to protracted negotiations on the allocation of risk between the two parties. Where the differences cannot be resolved, the negotiations may break down and the deal may be aborted.

Reduced sale price

The relative negotiating strength of vendor and purchaser is influenced by the quality and quantity of information each has available on the condition of the land and their resolve to use that information to defend or reduce the price. A vendor's position may be weakened if a purchaser provides evidence that the site is contaminated. If the sale is close to completion and the vendor cannot afford a further delay investigating the site for itself, it may have no option but to accept a reduced price and/or concede onerous indemnities.

Contingent liabilities (post-transaction)

Following the transaction, the vendor may retain some liability as a result of explicit or implicit (eg through omissions) conditions in the contract, or through legislation such as Part IIA EPA 1990.

6.2.2 The purchaser's financial risk exposure

Land contamination presents the following financial risks to purchasers:

- protracted or abortive negotiations
- impact of contamination on business performance
- acquisition of unforeseen liabilities.

Protracted or abortive negotiations

Known or suspected contamination can delay the completion of a transaction, resulting in additional expense or the abandonment of a deal.

Impact on business performance

A purchaser's business may be affected if contamination proves to be more significant than the pre-acquisition assessment suggests. In an extreme case, a site may turn out to be unfit for its intended purpose. The associated damage to a corporate or brand reputation can also be significant, even if the consequences of the contamination are not.

Acquisition of unforeseen liabilities

The *caveat emptor* maxim (see Box 6.1) can leave the purchaser with third party and regulatory liabilities.

Box 6.1 *Caveat emptor (adapted from Waite and Jewell, 1997)*

> *Caveat emptor* [Latin for "let the buyer beware"] is a common-law maxim warning a purchaser that he cannot claim that the land that he had purchased was defective unless he protected himself by obtaining express guarantees from the vendor. It will not protect a vendor who takes active steps to conceal a material fact about the property or who misdescribes or misrepresents a material fact about the property (Laidler et al, unpublished). Statutory conditions introduced by the Sale of Goods Act 1979 do not apply to sales of land, to which *caveat emptor* still applies as far as the condition of the property is concerned. However, a term is normally implied that the vendor must convey a good title to the land, free from encumbrances that were not disclosed to the purchaser before the contract was made.
>
> *Caveat emptor* only affects the contractual position between vendor and purchaser. It does not operate to transfer liabilities from the vendor to the purchaser, nor to relieve the vendor of liability. Nor can a purported transfer of liabilities from vendor to purchaser in the contract alter this position. It follows that the vendor retains liability for statutory clean-up and damages caused to third parties after the transaction. However, the vendor's liabilities may become diluted as subsequent purchasers acquire their own liabilities which run concurrently with those of the vendor. A vendor's statutory liability under Part IIA EPA 1990 may also be transferred to the purchaser through exclusion tests. These tests do not apply to common law actions or other statutory liabilities, and their effectiveness to the vendor depends on a number of conditions being satisfied (notably that the purchaser lies within the same liability group as the vendor).

6.3 MANAGEMENT OPTIONS

The following sections outline risk management options for vendors, purchasers, landlords and tenants.

6.3.1 The vendor's options

The main risk management objectives of the vendor are firstly to minimise the impact of contamination on the progress of the transaction and the price of the property/ company being sold, and secondly to manage the risk of retaining post-transaction liabilities. The following techniques can be used singularly or in combination to help meet these objectives:

- due diligence (loss prevention)
- restructure transaction to exclude sites (risk avoidance)
- vet the purchaser (loss prevention/loss mitigation)
- undertake remediation prior to disposal (loss prevention)
- retain rights to control future remediation (loss mitigation)
- position the company so that it falls within one or more of the Part IIA EPA 1990 exclusion tests (loss prevention/loss mitigation)
- transfer risk contractually (financial risk transfer)
- arrange prospective or retrospective financing structures (retained risk financing)
- set up composite risk financing programme (hybrids)
- insure risks (financial risk transfer).

Due diligence

Due diligence is designed to provide a vendor with greater certainty of the risks associated with a site, enabling the company to develop an appropriate sale strategy.

Information gathered during the due diligence process has considerable value to the vendor in terms of:

- providing a basis by which to challenge a purchaser's attempt to obtain a price reduction on environmental grounds
- providing a baseline survey which may be used at a later date as evidence in any disputes regarding the causation of contaminant-related damage at the site.

Restructure transaction

If a vendor is involved in a multi-site transaction, those contaminated sites which are perceived to present the greatest risk may be excluded from the sale. This ensures that control of the risk remains with the vendor. If the perceived risk is considered to be much greater than the actual risk, the sale price may not represent a satisfactory realisation of the vendor's assets. In such cases, delaying the sale until the perception of risk has altered, or economic factors have counteracted the stigma effect of contamination, may secure a more realistic price for the vendor.

Vet the purchaser

A vendor can reduce its risk exposure by selling the property to a responsible organisation with a proven record of effective environment management. Such companies may be more likely to manage any residual contamination in a responsible manner that avoids escalation of the risk. If the vendor is relying on counter indemnities to allocate risk, the creditworthiness of the purchaser is also an important factor that should be checked.

Remediation prior to disposal

This approach is likely to provide the vendor with greater protection in respect of future liabilities than an approach based on selling the site "contaminated". Remediation may also make the property more marketable and increase its value. The main limitation lies in the uncertainty surrounding the effectiveness of remediation, with the possibility that residual contamination may still be present beneath the site. The remediation process can, itself, also expose a company to financial risk, as discussed in Chapter 7. Time and technical constraints may preclude treatment, and in certain cases, the costs of remediation may not be fully recovered in terms of increased asset value.

If remediation is undertaken, it is prudent for the vendor to provide the purchaser with documentation detailing the nature of the work and the final condition of the land. This can be used at a later date to resolve disputes regarding responsibility for contamination at the site.

Retain rights to control remediation in the future

A vendor may wish to consider holding the right to manage any remediation should it retain liability for contamination. Care should be taken, however, not to step outside the exclusion test for "payments for remediation" by exercising too extensive rights of control over the land.

Part IIA EPA 1990 exclusion tests

The explicit strategy of avoiding liability under Part IIA by means of the exclusion tests requires at least one other party to be in the Class A liability group, and is fraught with uncertainty as to how regulators, and ultimately the courts, interpret the exclusion tests. Nevertheless, Part IIA liabilities may be transferred to a purchaser if the vendor can exclude itself from liability by virtue of satisfying the conditions in one or more of the exclusion tests, notably making payments for remediation and/or selling the site with information.

Other liabilities, such as civil actions by third parties, would not be covered by this approach.

Contractual risk transfer

Warranties and indemnities are usually the most important provisions of sale and purchase agreements in respect of apportioning environmental risk between vendors and purchasers.

A vendor may seek indemnities from a purchaser for losses arising from any one or more of the following (Waite and Jewell, 1997):

- contamination from carrying on the vendor's business of which the vendor is aware
- contamination from carrying on the vendor's business of which the vendor is unaware
- historic contamination of which the vendor is aware
- historic contamination of which the vendor is unaware
- contamination caused or exacerbated by the activities of the purchaser after completion (in particular creating new pathways, introducing new receptors or causing chemical alteration of pre-existing substances).

The market will rarely bear the level of indemnification sought by vendors – ideally unconditional indemnities provided by the purchaser for new and residual contamination with no counter-indemnities or warranties provided by the vendor in return. In addition, there are limitations associated with indemnities including:

- drafting and negotiation can be problematic and time consuming, since there are a number of issues to be addressed including, but not limited to:
  - whether the indemnity applies to breaches of new legislation
  - whether the indemnity applies to all contaminating substances or specified substances only
  - minimum (*de minimis*) value of claims that can be made under a warranty
  - whether consequential loss is included
  - the claim period
  - the standard of remediation
  - the events that trigger the indemnity (eg receipt of a statutory notice or voluntary remediation to avoid the serving of such a notice)
- indemnities claiming to cover criminal liability are not enforceable in UK courts
- indemnities may be for a limited period and subject to financial caps and other conditions
- the security of indemnities depends on the financial standing of the company providing them.

Purchasers will often seek warranties or indemnities from vendors. Whilst these should be resisted, there are situations where provision of warranties or indemnities is unavoidable. In such cases, vendors should be aware of the following pitfalls in negotiating and drafting warranties (John, 1999).

1. Non-environmental warranties may unwittingly provide a purchaser with warranty cover for contamination liabilities. Where specific environmental warranties are included, the vendor should ensure that the contract provides that these are "ring fenced", ie the only warranties given in relation to environmental matters.
2. Warranties that a site is not contaminated will mean that a vendor may face claims for damages arising from contamination that has migrated onto a vendor's site during its period of ownership, as well as from historic contamination that the vendor may or may not have caused. For this reason, from a vendor's perspective, warranties that the vendor has not contaminated its own property are preferable.
3. Warranties are often qualified by means of a disclosure letter which describes why, in some cases, certain facts of the warranty are not true. There is a standard clause that normally appears towards the front of a sale contract that states where warranties are qualified in this way they are deemed to be given by the vendor after making due and careful enquiry. Without amendment in the environmental section, this could imply that a vendor has made full enquiry of all regulatory authorities and conducted site investigation. If this is not the case, the appropriate warranties should be specifically amended.

The use of warranties and indemnities in contracts for the sale of property is discussed by Laidler et al (unpublished).

Retained risk financing

Retained risk financing involves building up funds prospectively or retrospectively to fund losses. This approach is suitable for all types of risk, but particularly those risks over which the company has little or no control. Expenditure is held within the company, hence there are financial benefits (compared to insurance) if risks do not give rise to significant losses. The main limitation of risk financing lies in the uncertainty of setting appropriate financing terms. If the terms are too low, the company will be faced with the risk of unplanned losses on the balance sheet, whilst setting terms too high will draw funds away from other areas of the business which may suffer as a result.

Composite risk financing

Risk financing can be blended with risk transfer as part of a composite structure. The approach is applicable to all contamination risks, and particularly those risks where losses are inevitable but the timing and cost of the loss is uncertain. Such risks may not be insurable. Composite structures offer budget stability for single or "bundled" risk exposures (see Chapter 10).

Insurance

Insurance can provide an effective mechanism to facilitate transactions and manage a vendor's contingent liabilities. It can also be used to reduce uncertainty in the effectiveness of other management techniques used during the transaction, for example, by insuring warranties and indemnities. A detailed review of insurance policies is included in Chapter 11.

6.3.2 The purchaser's options

The techniques to manage the financial risk of asset and share purchases are essentially the same, although more care needs to be exercised for share purchases to reflect the greater level of risk of acquiring liabilities.

Risk management options comprise:

- due diligence
- restructure transaction (risk avoidance/loss prevention/controlled risk transfer)
- position the company so that it falls within one or more of the Part IIA EPA 1990 exclusion tests (loss prevention/loss mitigation)
- transfer risk contractually (financial risk transfer)
- develop funds to manage the risk (retained risk financing)
- set up composite risk financing structures (hybrid risk financing)
- insure risks (financial risk transfer).

Due diligence

Due diligence reduces the uncertainty of a purchaser acquiring undisclosed liabilities or an impaired land asset for which the company may have no redress under *caveat emptor*. Information collected during the due diligence provides the basis to negotiate contractual terms or a price reduction that reflects the level of risk associated with the site.

The possibility of contamination should be investigated before contracts are exchanged, unless conditional contracts are agreed subject to a satisfactory site report. If a purchaser relies on a report commissioned by the vendor, the purchaser should check that the company conducting the survey/report owes it a proper duty of care. If existing reports are not available or satisfactory, the prospective buyer should consider commissioning a new report.

In the case of share purchases, the due diligence should check for residual liabilities associated with former ownership of properties (see Section 5.3.2). It is in the purchaser's interest to seek useful responses to all environmental enquiries.

The main limitations of due diligence are:

- the process may fail to identify, or underestimate the extent of, liabilities, irrespective of any negligence on the part of a consultant
- under certain circumstances, a vendor may be excluded from Part IIA EPA 1990 liability by virtue of the sold with information test
- information may not always be readily available, for example, in the case of a hostile take over
- time constraints may preclude comprehensive due diligence.

If the due diligence highlights potential or actual contamination liabilities, the purchaser has the following options:

- exclude site from transaction, if the site is part of a portfolio
- restructure share to asset purchase to increase the prospect of liability relating to past transgressions remaining with the vendor
- insist site is remediated by vendor
- agree a retention from the purchase price to cover cost of remediation, noting that in so doing, the purchaser takes on the vendor's Part IIA liabilities (assuming the purchaser was or subsequently became a Class A appropriate person for the relevant significant pollutant linkage)
- arrange insurance, if the risk is insurable
- seek indemnities from the vendor for losses arising from the identified contamination.

The following techniques can be used to manage a purchaser's risk of acquiring liabilities which were undetected or underestimated during the due diligence.

Contractual risk transfer

Warranties and indemnities provide an important contractual mechanism for allocating risk between vendor and purchaser. The extent of environmental warranties will depend on the nature of a target's business – less warranty cover will be required for a "people" business than for a heavy industrial business (Phillips and Ferera, 1998). Depending on the relative bargaining strengths of the parties, a purchaser can attempt to obtain warranties as to the veracity of certain matters, including:

- the vendor has complied with laws
- the vendor has complied with all licences and consents
- the vendor has not undertaken any activity on the site which is likely to cause or exacerbate contamination
- the vendor has remediated the site to a give standard
- the vendor knows of no circumstances which will or may give rise to liabilities under environmental law
- the site is not contaminated.

Warranties require the claimant to establish the damage or loss suffered before it becomes payable. For this reason, indemnities are preferable to warranties (yet more difficult to secure), since a claimant under an indemnity only needs to show that the event that caused the loss has occurred for the amount indemnified to become payable.

Depending on its negotiation strength, a purchaser may seek to obtain indemnities from the vendor in respect of liabilities for pre-completion incidents, which subsequently attach to the purchaser.

If the vendor is merely a holding company or, in the case of a business, will be merely a shell after completion of the transaction, a purchaser may find it has a claim against a vendor that lacks sufficient assets to meet the claim. In such cases, a purchaser should consider seeking a parent company or other guarantee. In addition, where a corporate seller's assets are largely in the form of distributable reserves, a purchaser should also consider imposing a restriction on the distribution of these reserves during the period of the warranty or indemnity (Phillips and Ferera, 1998).

Alternatively, specialist insurance policies may protect the security of a warranty or indemnity, or provide an alternative (and potentially more secure) method to transfer risk (see Chapter 11).

Retained risk financing and composite structures

A purchaser may wish to protect its investment by building up funds to pay for losses. Funding arrangements may also be blended with risk transfer mechanisms as part of a composite risk financing strategy (see Chapter 10).

Insurance

Insurance provides an effective mechanism to facilitate transactions and manage the purchaser's risks of incurring liabilities and of allowing for a compromise between the vendor's and purchaser's positions with respect to contractual risk allocation. A detailed review of insurance products is included in Chapter 11.

6.3.3 The landlord's options

The management objectives of a prospective landlord include minimising the risk of incurring liabilities as a result of the tenant's future activities, and managing the effects of contamination on income received through the lease. The techniques described in Section 6.3.1 can be used singularly or in combination to help meet these objectives. Landlords may also consider negotiating the inclusion of the following lease provisions.

1. Restriction of use of property: this can be used to prevent a polluting activity. However, the more restrictive the clause, the more it will have a detrimental effect on the rent review and thus the value of the lease inclusion.
2. Covenant to comply with statutory obligations: landlords should seek to ensure that tenants meet all statutory obligations during their occupancy of a site, but should be aware that many tenants, if properly advised, will refuse to accept such obligations in relation to contamination which pre dates the lease.
3. Covenant to comply with good industrial practices: covenants such as these, which may require that the tenant complies with relevant codes of practice, can reduce the likelihood of new releases of contamination occurring as a result of poor environmental management.
4. Covenant against nuisance or annoyances: a lease may contain a covenant that the tenant will not bring anything onto the demised property likely to cause a nuisance (in particular harmful substances), including an indemnity to the landlord against so doing. Annoyance is a wider concept for which breach of the covenant will be easier to prove since the criteria for nuisance does not have to be established.
5. Repairing covenants and yielding up obligations: repairing covenants do not usually provide a reliable mechanism for achieving remedial action unless contamination affects a building, including the services. Exceptions include covenants which expressly include remediation or which are worded in such broad terms to cover all defects irrespective of whether actual physical damage is caused.

A covenant to "keep in good condition" goes beyond mere repair and may involve the obligation to remediate contaminated soil within the property, so as to satisfy the following requirements:

- ensuring safety of people working, living or visiting the property for the purpose of the permitted use
- preventing pollutants migrating from the property so as to give rise to liabilities on the owners and/or occupiers.

6. Rights of entry: a covenant to allow the landlord to enter the property to assess ground contamination, undertake remedial action and recover the costs from the tenant will provide the landlord with control over the risk. This brings with it the potential for the landlord to be held liable under Part IIA as a causer or knowing permitter.

A baseline survey should be conducted at the beginning of a lease to help differentiate between contamination originating before and after the lease commenced. In all cases involving the negotiation and drafting of contractual provisions, specialist legal advice should be sought.

6.3.4 The tenant's options

The management objectives of a tenant entering into a leasehold agreement include minimising the risk of incurring liabilities for historic conditions that pre-date the tenancy, and managing any adverse effects of contamination on its business. The techniques described in Section 6.3.2 can be used singularly or in combination to help meet these objectives.

Where possible a tenant should resist the lease terms described in Section 6.3.3, whilst considering the following additional protective measures (Waite and Jewell, 1997).

Piecrust lease

Under a pie crust lease, the property is limited to the space from ground level upwards. By this means, a tenant does not occupy the underlying land and should, therefore, avoid liability for any contamination unless the tenant causes it or unless contamination migrates on to land which is within the demised premises. This type of lease is mainly appropriate for short-term leases. It may also be unsuitable for tenants planning to develop the site since there are likely to be restricted rights to dig foundations.

Lease provisions

Tenants may seek to negotiate the following lease provisions.

1. A covenant by the landlord to remediate any contamination on the demised premises at the commencement of the lease (if due diligence shows the site to be contaminated).
2. A covenant by the landlord to take action against tenants of other parts of the property in respect of pollution arising from those premises and (if unlet) to deal with any fly-tipping/squatters.
3. An indemnity from the landlord in respect of any contamination liabilities caused prior to the commencement of the lease.
4. A suspension of rent and service charge during any period in which remedial action carried out on the property in respect of pre-lease contamination causes unreasonable disruption to the tenant.

5. Break clauses enabling the tenant to surrender the lease if pre-lease contamination is discovered on the property which is likely to give rise to material liabilities on the tenant or to give rise to unreasonable disturbance to his business.
6. A service charge that does not allow the landlord to recover costs of past contamination from the tenant.

Note: To be fully effective, covenants and indemnities relating to the condition of the site should be accompanied by a baseline survey at the commencement of the lease. This can help differentiate between contamination present before and after the lease commenced, although differentiation may still prove to be difficult.

In all cases involving the negotiation and drafting of contractual provisions, specialist legal advice should be sought.

7 Development and contracting

This chapter identifies the contamination-related financial risks associated with land development and contracting (including remediation), together with options to manage these risks. Chapter 4 (Section 4.2.5) presents an approach for using this information to develop an appropriate risk management strategy.

7.1 BACKGROUND

The planning regime provides the principal mechanism for remediating historic contamination in the UK. There have been a number of cases where the discovery of contamination during development has led to major delays and extra costs (Mallett, 1999).

7.1.1 Stakeholders

The stakeholders in development and contracting comprise:

- risk holders: developers and site purchasers
- potential holders of primary liability: contracting engineers; remediation technology vendors; financiers; investors and insurers
- other third parties: professional advisers; regulators and the general public.

7.2 RISK EXPOSURES

The financial risk exposures of developers and contractors are summarised in Table 7.1 and discussed further in Sections 7.2.1 and 7.2.2 respectively.

7.2.1 Developer's financial risk exposure

Land contamination presents the following financial risks to developers:

- adverse impact on the financial performance of a project
- liabilities incurred during the development
- contingent liabilities post-development.

The circumstances by which these risks materialise are discussed below.

Adverse impact on financial performance of a project

The discovery of historic contamination may prevent a development project proceeding beyond due diligence, causing the developer to lose the money it invested up to that stage. More damaging financially in terms of wasted expenditure is the discovery of unforeseen historic contamination during the development itself. The extra costs needed to manage this contamination may be so high, or the feasibility of remediation so low, to preclude the intended development. In such cases, the project may be aborted, sold on at a loss, or the site developed for a less sensitive (and potentially less valuable) use.

Table 7.1 *Financial risk exposure of developers and contractors*

Financial impact (if risk materialises) / **Financial risks**	**Wasted expenditure** [1]	**Loss of profit**	**Additional project costs** [2]	**Asset impairment** [3]	**Financial penalties** [4]	**Contractual liabilities**	**Enforced remediation** [5]	**Third party liabilities** [6]	**Legal costs**	**Fines** [7]	**Damage to reputation**	**Key factors affecting the risk**
Developer												
Adverse financial performance of development, including aborted project	✓	✓	✓	✓	✓						✓	Uncertainty of the effects of known and unknown contamination on the project pricing, timing, viability and ease of securing planning permission Uncertainty in the capacity of project to bear extra unforeseen costs and liabilities
Liabilities (during development)							✓	✓	✓	✓	✓	Uncertainty of the effects of contractors' actions in relation to exacerbating existing contamination or generating new contamination
Contingent liabilities (post sale)					✓		✓	✓	✓			Uncertainty of future financial impact of, and responsibility for, residual contamination [8]
Contractor												
Liabilities (not fully taken into account in contract price)		✓	✓		✓	✓	✓	✓	✓	✓	✓	Uncertainties of the risks presented by undetected contamination in terms of third party liabilities for exacerbation of contamination; contractual obligations to undertake extra un-priced works or pay for liquidated damages.
Adverse financial performance on contract		✓	✓		✓						✓	Uncertainties of the risks of undetected contamination increasing project costs
Contingent liabilities (as a result of providing warranties/ indemnities)					✓							Uncertainty of future financial impact of residual contamination and the effectiveness of the remediation/ development process

Notes

1. Eg wasted investment occurring as a result of an aborted project.
2. Eg extra costs to a) make the site fit for purpose; b) reduce the risk of incurring future liabilities; c) undertake long term monitoring, and d) secure additional finance and/or insurance.
3. Eg increased asset realisation period, reduced value of the asset, inability to generate income.
4. Eg penalties for delaying on or failing to deliver asset to purchaser.
5. Part IIA EPA 1990, WRA 1991.
6. In respect of bodily injury and property damage (employees, visitors, neighbours are potential claimants).
7. Including actions pertaining to health and safety under the CDM Regulations 1994.
8. Eg future hazards posed by contamination may change as a result of changes to legislation, introduction of new pathways or receptors.

The extra costs to manage contamination can derive from any or all of the following:

- soil and groundwater treatment
- disposal of contaminated soils, including landfill charges, haulage and replacement of excavated material
- refinancing of loan.

Commercial land transactions may proceed by way of conditional contracts, which require the purchaser to complete on the deal upon the happening of a specified event. Alternatively, contracts may contain options to purchase, whereby a purchaser can call for the transfer of freehold upon the happening of an event, such as the granting of a satisfactory planning permission. Both forms of contract can have significant cost implications to a developer faced by known or suspected contamination. For example, a purchaser under an option to purchase may covenant to use its best endeavours to obtain a satisfactory planning permission. The company may then be forced to undertake expensive investigations required by a local authority as part of the planning process. Similarly, a purchaser that signs up to a conditional contract for which the trigger is a "granting of a specified planning permission" may be compelled to complete on a deal which only provides outline planning permission subject to a site investigation (Graham, 1995).

Discovery of unforeseen contamination can also have an effect on the project timing, which may in turn result in additional costs and liabilities. For example, where the development has secured a pre-let, delays in completing the project can have a knock-on effect in terms of preventing the timely hand over of the asset. The developer will lose its projected rental income and, depending on the nature of the lease, may incur financial penalties or give the prospective tenant the right to walk away from the deal.

Contamination may also impact on the value of the property, even if remediation is undertaken to the satisfaction of all concerned. This reflects the potential stigma associated with contamination, together with the relatively weak negotiation position of a developer.

Liabilities

The development process may exacerbate historic contamination or cause a new release. Either could trigger third party liabilities in respect of bodily injury or property damage, enforced remediation under Part IIA EPA 1990 or WRA 1991 and/or action by the HSE. A developer should not normally be held liable for pollution caused by the actions of a contractor. However in the case of potential for bodily injury, developers should note that their developments will probably come under the scope of the Construction (Design and Management) Regulations 1994 (also known as the CDM Regulations 1994). As such the developer is responsible for appoint a Planning Supervisor to advise on health and safety aspects of the proposed work. Failure to appoint a planning supervisor could render a developer liable to prosecution. Advice to clients on CDM Regulations is provided in CIRIA's publication, *CDM Regulations – Practical guidance for clients and clients' agents* (prepared by WS Atkins Consultant Ltd. 1998).

Following the sale of the site, residual contamination can leave a developer vulnerable to third party liability claims arising from exposure to the contaminants, or enforced remediation as a result of new, more stringent, legislation.

7.2.2 Contractor's financial risk exposure

A contractor's financial risk exposure comprises:

- liabilities for causing pollution during the development process
- contractual liability for extra project costs and/or liquidated damages (depending upon the nature of its contract)
- contractual liabilities following development.

A contractor's exposure to these risks will vary according to the nature of the construction contract into which it enters. Companies engaged in fixed-price contracts are likely to be most vulnerable to claims arising out of pollution caused during the development. Private finance initiative (PFI) projects, where a contractor's involvement in a project may continue long after the construction works have been completed, may expose it to environmental liabilities arising out of maintenance or commissioning works, or operation of the facility in question. (Stubbs, 1998). A contractor entering into a collateral warranty may increase its risk exposure, since the warranty gives another party the right to sue for losses arising from breach of contract. During the process of resolving disputes with a client, a contractor may also suffer impact on cash flow.

7.3 MANAGEMENT OPTIONS

The principal management objectives of developers and contractors are to manage the risk of incurring liabilities and extra project costs, or suffering reduced income. The following sections outline options for both groups of organisations to manage these financial risks with respect to contamination.

7.3.1 Developer's and contractor's options

The following risk management options are available to developers and contractors:

- project due diligence (loss prevention)
- remediation (loss prevention)
- contracting (controlled risk transfer)
- warranties (financial risk transfer)
- parent company guarantees and bonds (loss mitigation)
- development of an appropriate exit strategy (loss mitigation and loss prevention)
- prospective and retrospective financing (retained risk financing)
- insurance (financial risk transfer)
- composite risk financing (hybrids).

Project due diligence

The potential financial impact, if any, of contamination should be assessed during the project feasibility stage. This enables the risk to be allocated and managed more effectively, including avoiding sites where appropriate. Project due diligence also allows the principal areas of financial uncertainty to be identified. However, the process is not infallible, and may still fail to detect contamination. As a result, developers and contractors should consider combining due diligence with other management measures, even if the due diligence fails to identify any material financial risk.

Remediation

If contamination is discovered before or during the development, from a risk management perspective, it is preferable that the contaminants are treated, removed or destroyed rather than contained on-site. Where removal to landfill is not commercially viable, it is advisable to use a non-reversible, "destructive" treatment process where possible. This reduces the risk of contingent liabilities.

Contracting

Contracting provides the principal mechanism for a developer to transfer part or all of its risk to another organisation. For unknown contamination, a developer may attempt to transfer the responsibility to pay for any additional costs to a contractor. In response, the contractor's price is likely to increase to pay for the costs necessary to assess uncertainty in the level of risk and manage the risk accordingly. Alternatively, a risk sharing arrangement may be agreed, whereby both parties subsidise the costs of any additional work and share any cost savings. Risk-sharing contracts can be structured to incorporate an insurance element, as described in Box 11.2 in Chapter 11.

Table 7.2 provides a summary of the allocation of risk for various types of contract, together with other characteristics that are typically considered during the selection of an appropriate form of contract. Whichever contract is selected, it is important that the contract clearly defines who is responsible for paying for any costs associated with unforeseen contamination. This can be achieved effectively through the use of express conditions including:

- penalties for late completion
- indemnification for liquidated damages
- fitness for purpose or reuse covenant
- covenant to maintain insurance.

Table 7.2 *Apportionment of risk for various types of contract*

	Contract type							
Contract feature	**Selection criteria**	**Traditional**	**Traditional measurement**	**Traditional cost plus forms**	**Traditional short forms**	**Design and build**	**Management**	**ECC target**
Ownership of risk	Developer retains risk						✓	
	Shared risk	✓	✓	✓	✓	✓	✓	✓
	Contractor holds risk					✓		
Timing	Crucial						✓	✓
	Important		✓	✓		✓		
	Less important that other factors	✓	✓	✓	✓			
Price certainty	Very important	✓						
	Important	✓	✓		✓	✓	✓	
	A target is acceptable			✓			✓	✓
Complexity	Complex design	✓					✓	
	Moderately complex design		✓	✓	✓	✓	✓	✓
	Simple design				✓			

Parent company guarantees and bonds

The effectiveness of any transfer of risk from developer to contractor depends on the financial security of the contractor. Parent company guarantees or secure bonds provide a means to recover costs in the event that a contractor fails to meet its contractual obligations and is not in a financial position to compensate the client.

Development of an appropriate sale strategy

In the case of contingent liabilities, which the developer may face following the sale of a site, the following steps can be followed to minimise the risk.

1. Provide full documentation of any remediation or site investigations in the conveyancing records. This will increase the possibility that a prospective purchaser will fall into the Class A liability group for contaminated land under Part IIA. The records may also form the basis of an exclusion from liability for contamination caused by future owners and occupiers.
2. Subject to negotiations, avoid, where possible, providing warranties (especially those relating to fitness for purpose) and indemnities to purchasers.
3. Seek, where possible, indemnities that clearly assign responsibility for new contamination arising post completion to the purchaser.
4. Finance potential losses by building up funds prospectively, or paying for losses retrospectively. Alternatively contingent liabilities may be financed externally through specialist insurance – see Chapter 11.

Chapter 6 provides more details on managing the risks of land transactions.

Prospective and retrospective risk financing

Retained risk financing involves building up funds prospectively or retrospectively to fund losses. The technique is applicable to all contamination losses, and is particularly relevant to uncontrollable risks such as contingent liabilities after sale of the site.

1. Structured project finance: the financing structure will define the extent to which a project can bear risks. For example debt finance towards the end of a project, combined with higher levels of equity financing towards the beginning of a project are more likely to be acceptable to lenders. Higher levels of debt at the beginning of the project could reduce the developer's risk exposure, although this structure is less likely to be a bankable financial proposal.
2. Fall-back financing options: in the event that the project is delayed as a result of contamination, pre-agreed fall-back financing options will help to limit the financial implications. Options may be as simple as agreeing additional credit lines with a lender or providing contingency reserves. More complicated arrangements can include providing equity finance at various stages in the project.

Insurance

Developers and contractors may wish to consider appropriate insurance cover (eg cost cap/remediation stop loss, third party liability, own site remediation, property transfer, legal indemnity or contractors all risk policies) for the development and post-development phases of a project. Insurance may be purchased directly or a developer may rely on cover provided by a contractor's policy. A detailed description of insurance is included in Chapter 11.

8 Finance and investment

This chapter identifies the contamination-related financial risks associated with investment and provision of finance, together with options to manage these risks. Chapter 4 (Section 4.2.5) presents an approach for using this information to develop an appropriate risk management strategy.

8.1 BACKGROUND

Finance and investment is needed to fund projects (eg redevelopment, construction, brownfield reclamation) and businesses (eg operating processes and commercial property transactions). The main types of finance are outlined in Box 8.1.

The source, terms and conditions of finance can have an important effect on the capacity of a project or business to bear financial loss.

For the purposes of this report, investment in property ownership (as shown by the company being named on the title deeds) is deemed to be a landowning activity which is covered in Chapter 5.

Box 8.1 *Selected types of finance (based on Hunt and Hilton, 1997)*

Secured short-term debt: a secured debt normally encompasses a first fixed or floating charge, mortgage or debenture over the property assets, possibly linked to additional security such as personal guarantees from directors or shareholders.

Unsecured short and medium-term debt: committed unsecured short-term and medium term loans are available, but banks are generally unwilling to lend for more than about five years. Examples include bilateral loans (direct loans between the borrower and lender) which take many forms relating to term, interest and principal payment structure, security and covenants.

Long term debt and debt securities: these financing arrangements are normally secured and offer borrowers a fixed rate of interest. They have the additional advantage that capital repayments are not usually due until maturity, matching the long-term nature of property assets. Securitisation is the process of converting rental income into bonds or other interest-paying IOUs that can be sold in the capital markets. Investors who buy these bonds generally comprise pension funds, life insurers and banking groups. The UK contributed to 90 per cent of European property-related securitisation in 1999 with the volume of UK deals topping £4.2 billion in 1999 (Catalano and Frampton, 2000).

Bond financing is a form of long-term finance, which does not require annual repayment and fixes interest costs throughout the duration of the fixed term.

Mezzanine finance is a form of subordinate debt, usually carrying rights to convert to equity in the future, which is used to fill the gap between equity and conventional debt finance.

Equity investment is normally not redeemable, and is dependent on the financial performance and profitability of the assets, which are in turn dependent on increases in capital value and the flow of income derived from it. Returns from equity investment are typically more variable, volatile and insecure than debt.

Joint ventures are designed to spread risk and give a share of profits between two or more companies participating in a project or venture.

8.1.1 Stakeholders

The stakeholders in the provision of finance and investment include:

- risk holders: lenders and investors
- third parties affecting the risk and liability of lenders and investors: borrowers
- other parties: professional advisers, regulators and general public.

8.2 RISK EXPOSURES

The financial risks that contamination presents to lenders and investors are summarised in Table 8.1 and discussed in Sections 8.2.1 and 8.2.2.

Table 8.1 *Financial risk exposure of lenders and investors*

Financial impact (if risk materialises) / **Financial risks**	**Failure to recover principal and/or interest on loan**	**Loss of initial investment (equity)**	**Reduction in investment return[1]**	**Decrease in shareholder value**	**Business disruption**	**Extra expenditure to realise value of security[2]**	**Tied up investment (equity)**	**Enforced remediation**	**Third party liability[3]**	**Fines and injunctions**	**Legal costs**	**Damage to corporate reputation**	**Key factors affecting risk**
Lender													
Loan default (*solvency risk*)	✓				✓								Uncertainty of the effect of contamination on the ability of a borrower to repay a loan
Adverse impact on value of security (*collateral risk*)	✓					✓							Uncertainty of the possible impact of contamination on the worth of assets as a security on a loan
Acquisition of liabilities	✓				✓	✓	✓	✓	✓	✓	✓	✓	Uncertainties surrounding the control to which a lender may exercise over a borrower's activities without the lender acquiring the liabilities of the borrower[4]
Investor													
Adverse financial performance of an investment		✓	✓	✓			✓					✓	Uncertainty of the effect of contamination on the performance of a project, venture or business

Notes

1. Eg income generation and/or capital growth on the investment.
2. Eg in the event of taking security of a loan, a lender may need to undertake remediation to make the site suitable for use and attract investment or a purchaser of the site.
3. In respect of bodily injury and property damage.
4. Enforcement of security by taking possession of a site can result in a lender acquiring a borrower's liabilities.

8.2.1 The lender's financial risk exposure

Lenders have three main concerns regarding the financial impact of land contamination:

- loan default (solvency risk)
- diminution in the value of assets which form security on a loan (collateral risk)
- acquisition of a borrower's liabilities (acquired liability risk)

Solvency risk

Lenders can be affected if contamination reduces the ability of a borrower to meet its repayment obligations. This can occur, for example, if a borrower suffers unforeseen extra expenditure or reduced income flow as a result of new or historic contamination. Such losses can affect the liquidity of the borrower, which in the worst case can lead to the borrower defaulting on a loan, putting the lender's principal and interest in jeopardy. A lender may also incur extra costs in managing the effects of the loan default.

Collateral risk

Contamination can impact on land value (even to the extent of causing negative value when the costs of necessary remedial measures are taken into account), reducing its worth as security for a loan and increasing the vulnerability of a lender in the event that a borrower defaults. In certain cases, claims made by regulatory authorities under Part IIA EPA 1990 for their costs of remediating the site can be made a charge on the premises. Being a statutory charge, this would probably take priority over a lender's mortgage or legal charge.

Acquired liabilities

A lender is only likely to incur liabilities if it exercises control over the polluting activities of its borrower or becomes a mortgagee in possession of the site. Simply loaning money should not trigger lender liability because the lender is not actively controlling the borrower's activities, whilst mortgagees not in possession are specifically excluded from owner/occupier liabilities under Part IIA EPA 1990. Thus, it follows that a lender should be mindful of its strategy during the pendency of a loan, and of any additional involvement it has in a borrower's affairs, especially where the involvement may be seen to give the lender power to instruct and direct the borrower.

8.2.2 The investor's financial risk exposure

Investors' concerns about contamination relate to the potential adverse effect on the performance of their investments (ie projects or businesses) including:

- reduced value of shareholding or stakeholding
- decreased investment return, eg dividends
- loss of initial investment.

Historic or new contamination can contribute to these financial risks by causing a business/project to suffer additional expenditure or reduced income. When a business or project is insufficiently capitalised to meet these costs, and a loan is not forthcoming, additional funding will be required from shareholders. If the funds are not provided, the business or project may become illiquid or insolvent. In extreme cases, the creditors of the business or project may have recourse to the courts to declare the enterprise insolvent and apply for appointment of a receiver or administrator.

In the event that a business is declared insolvent, the receiver or administrator has responsibility to ensure that secured creditors are repaid, followed by trade creditors. Depending on the nature of an investor's financial commitment to the company, investors may fall behind secured and trade creditors for repayment. Thus, investors who take a shareholding in a business or project are in a primary loss-taking position.

Even if there is not a direct requirement by an investor to address land contamination liabilities, a financial loss may still occur through the value of the company diminishing.

Investor liabilities

Shareholders will only incur liabilities up to the value of their investment, irrespective of whether they have a controlling or minority interest. Those with a controlling interest may be held personally liable.

8.3 MANAGEMENT OPTIONS

The following sections outline options for lenders and investors to manage the financial risks presented by contamination.

8.3.1 The lender's options

The main objectives of a lender's risk management programme are to manage the risks of a borrower defaulting on a loan and the lender acquiring liabilities. As a last resort, a lender will seek to manage the risks associated with the enforcement of its security. The following risk management options are available to lenders:

- undertake loan appraisal (risk control/information management)
- avoid loan (risk avoidance)
- reduce the level of costs that are supported by a loan (loss reduction)
- set protective loan terms (loss mitigation)
- impose loan conditions (loss prevention and/or loss mitigation)
- refinance loan (loss prevention and/or loss mitigation)
- structure a syndicate loan (loss mitigation)
- enforce security (loss prevention/mitigation)
- insure risk (financial risk transfer)
- set aside provisions for bad debt (retained risk financing)
- set up composite risk financing structures (hybrid).

Loan appraisal

The purpose of the loan appraisal process is to assess the strength of a deal in terms of the ability of a borrower to repay a loan in full (including interest), the value of security, and the lender's risk of acquiring liabilities in the event of a default. The appraisal should encompass all elements of the borrower's proposal which may adversely affect the deal.

The factors outlined in Table 8.2 can be used to assess the potential effects of contamination.

Table 8.2 *Criteria for assessing the viability of a project*

Risk factor	Comments
Nature of the business or project	Influences the risk of a borrower causing pollution. Although the risk is highly project-specific, a built project will generally expose a lender to less risk than a brownfield site purchase/reclamation or a redevelopment project. Lending on a potentially polluting business will tend to expose a lender to greater risk than lending on one with relatively innocuous operations. However, even non-polluting businesses may be adversely affected by contamination, for example, by the migration of contaminants on to the property from surrounding land.
Track record of borrower, including its compliance record, commitment to environmental management, environmental awareness	Indicates the experience of successfully managing the risks on previous projects, although this is not necessarily an accurate indicator of future commitment to environmental management by the borrower.
Location of property in respect of the relative supply and demand	Strong demand may counteract the effect of contamination on the value of the property/business. This will change with time and is subject to considerable geographical variation.
Level of project due diligence which the borrower has undertaken, and appropriateness and completeness of the corresponding risk management	Demonstrates that the borrower is aware of, and has considered, contamination risks. Depending on the quality of the due diligence, the lender may rely on the findings, subject to the associated limitations and risks discussed in Chapter 9.
Project financing structure	The financing structure will define the extent to which a project can bear risks. For example, debt financing towards the beginning of a project, followed by an injection of equity towards the end can provide an effective financial risk alleviation mechanism.
Insurance	Appropriate insurance can reduce a lender's risk exposure.

Loan avoidance

The loan appraisal process acts to reduce uncertainty in the lender's risk exposure, enabling it to make more informed decisions on the financial viability of the proposed project. For commercial reasons, lenders will rarely avoid a loan, seeking instead to arrive at a solution that meets its own requirements and those of its clients. This may involve obtaining additional evidence to satisfy a lender's concerns and/or implementing any, or all, of the following measures to manage the risk.

Loan structure, terms and conditions

Repayment criteria can be set according to the financial risks perceived by the lender (factoring in contamination and all other risks for the project as a whole). Examples of more stringent loan terms include requiring higher levels of borrower equity, higher levels of security and increasing the loan life:cover and loan:value ratios.

An appropriate funding structure provides sufficient equity in a borrower's business to enable it to survive in the near term and sufficient debt finance to cover fixed and working capital requirements. The term and structure of debt being offered should match the assets being financed. Generally speaking, fixed assets such as property should be financed with long term funding such as term loans, commercial mortgages or finance leases (Hunt and Hilton, 1997).

A lender may consider protecting its financial position by including provisions in the loan documentation. These provisions can take the form of specific representations and warranties of compliance, covenants to maintain compliance and indemnities against any loss that the lender may suffer as a result of non-compliance. Examples of loan conditions relating to contamination risks include the following (Waite and Jewell, 1997).

1. Warranties:
   - property is free from hazardous substances
   - no environmental claims have been brought against the borrower
   - borrower is unaware of any circumstance which is likely to give rise to any environmental claim
   - borrower is in full compliance with environmental law applicable to its business and that it has obtained all necessary environmental consents, licences and other permits.

2. Covenants:
   - to comply with environmental law
   - to maintain in force all environmental licences, consents and other permits to carry on the borrower's business and to comply with any conditions attached thereto
   - to provide to the lender on request all information relating to the borrower's compliance with environmental law, including conditions attached to any environmental licence, consent or other permit
   - to maintain environmental insurance provided that it is obtainable on reasonable terms
   - to comply with environmental best practice
   - to ensure that the site is kept free from hazardous substances.

3. Conditions:
   - an environmental report is produced for the lender confirming that (a) operations on a property comply with environmental law in all respects and (b) there are no material liabilities relating to the condition of the property
   - lender is indemnified for any remediation costs which may fall on it and for any loss that it may suffer as a result of non-compliance with environmental laws
   - provision for early repayment in the event of any breach of environmental law or of any covenant by the borrower
   - provision for early repayment in the event of any change in environmental law which gives rise to a material risk of liability on the lender in respect of the property, the business of the borrower, or which may cause the rights of any third party to rank ahead of those of the lender.

As a result of conditions in the loan, the lender may be able to foreclose early, before the borrower's activities actually trigger liability or adversely affect its financial performance. Other conditions are designed to reduce the risk of a borrower creating liabilities through poor environmental management practices.

Whilst the inclusion of conditions may provide the lender with confidence that a borrower is undertaking its business activities in a manner which reduces the risk of creating contamination liabilities, the conditions mean that the lender may effectively take a controlling interest in the borrower's operations. Under certain circumstances, it may, therefore, acquire liabilities as a causer or knowing permitter under Part IIA. For example, if a lender does not call its default and enforce its security following a breach of an environmental covenant, it may be conceivably regarded as a knowing permitter of the continuing contamination, because it knows of the problem, has the power to prevent or reduce it and does not exercise that power.

Refinancing loans

In the event that a borrower suffers unexpected losses, refinancing can be used to help the business recover. Before proceeding with any refinancing, the lender should reappraise the project or company to assess whether it is "throwing good money after bad".

Syndicate loans

Lenders may reduce their potential risk exposure by providing syndicate loans which involve funding from more than one financial institution.

Enforce security

Before taking any steps to enforce its security, a lender should assess whether it is likely to incur environmental liabilities in so doing. Taking possession is probably the least attractive option since a lender is exposed to significantly increased risk (see Section 8.2.1) by virtue of it assuming ownership/occupancy status. The appointment of a receiver (which acts on behalf of the borrower) is preferable, since it avoids a lender having to take possession and control operations. However, a lender should be mindful of the receiver being seen to act as its agent, which may increase the lender's control over the operations, and hence expose it to risks as a causer/knowing permitter.

Insurance

Insurance is available to lenders to protect their financial interests. Relevant policies are described in Chapter 11.

Bad debt provisions

Lenders may consider making provisions for bad debts, particularly for loans where it is uncertain if the principal or capitalised interest will be repaid. In some cases, a write-down of the loan may give rise to a reduced cost compared to foreclosure and enforcement of security. If the lender opts for bad debt provision, it may consider securing its position by swapping some of its loan for equity ie a debt–equity swap.

Composite risk financing

Composite risk financing, such as ART, is a blend of risk transfer (eg insurance) and funding. For lenders, composite structures may be used to secure a loan, or improve its terms. ART might involve the diversion of provisions from the project into an earmarked fund to deal with the potential accrual of land contamination liabilities, prior to repayment of the loan.

8.3.2 The investor's options

Management options for investors are limited to:

- undertaking investment appraisal (risk control/information management)
- avoiding investment (risk avoidance)
- selling investment (loss mitigation).

Undertake investment appraisal

Typically, decisions to invest in a project or business are based on a) an appraisal of the forecast returns from a prospective investment and the associated financial risks over the projected duration of the investment, and b) a comparison of the appraisal findings with those from alternative investments. Land contamination risks should be factored into appraisals.

If the appraisal indicates the returns to be too low, the investor can either negotiate on the returns from the project, or opt to invest in alternative opportunities. If the risks are too high, then the investor will need to seek amendments to the risk-taking structure to reduce the risk relative to the returns. This might be achieved financially or by changing the project structure.

Sell investment

The investment can be disposed of partially or wholesale. Partial disposal of the investment to permit purchase of the investments might allow the risk to be diversified through portfolio holdings.

9 Professional advice

This chapter identifies the financial risks of providing or relying on professional advice on contamination-related matters, and highlights options to manage these risks. Chapter 4 (Section 4.2.5) presents an approach for using this information to develop an appropriate risk management strategy.

9.1 BACKGROUND

Professional advice is needed for all of the activities covered in Chapters 5–8, and may cover legal, valuation or technical aspects of risk management.

9.1.1 Stakeholders

The stakeholders in risks associated with providing and relying on professional advice comprise:

- primary risk holders: client organisations, PI insurers
- potential holders of liability: lawyers, contaminated land consultants, chartered surveyors, laboratories etc
- other parties: regulators, subcontractors, general public.

It should be noted that a significant number of professional advisers with specialisms in land now offer services in land contamination. Clients are advised to ensure that they are commissioning work from competent land contamination consultants by following the guidance in Contaminated Land Research (CLR) report 12 [DETR, 1999a].

9.2 RISK EXPOSURES

The sources of risk and the nature of potential financial losses for advisers and their clients are summarised in Table 9.1.

9.2.1 Adviser's financial risk exposure

An adviser's main source of financial risk arises out of differences between what the client wants and what the adviser can deliver. This gap may reflect natural uncertainty or uncertainty arising out of the limited resources the client has made available to the adviser. There is also the potential for an adviser to promise too much at tender stage, leaving it open to subsequent financial penalties, which in certain cases may be unreasonable in comparison to the overall fee value.

Project liabilities

The terms of engagement specify an adviser's role in, and contribution to, a project, and thus influence an adviser's exposure to liabilities. An adviser should examine in detail the terms of engagement, focusing particularly on any conditions that could give rise to liquidated damages or contractual penalties.

Organisational liabilities

Advisers can suffer loss at an organisational level if problems develop on individual projects. Corporate reputation can be damaged if an adviser's actions or omissions generate adverse publicity. Organisations can be fined and their officers imprisoned if serious breaches of environmental or health and safety legislation take place. Defending claims can result in substantial legal fees and divert senior management from profit-making business activities.

Table 9.1 *Financial risk exposure of advisers and their clients*

Financial impact (if risk materialises) / Financial risks	Escalating project costs	Liquidated damages	Contractual penalties	Legal costs	Fines	Damage to reputation	Key risk factors
Adviser							
Project-related liabilities	✓	✓	✓				Uncertainties surrounding the: – terms of contract – nature and extent of "duty of care"
Organisation-related liabilities				✓	✓	✓	Poor working practices giving rise to, for example. prosecutions by HSE for breach of health and safety legislation.
Client							
Increased liabilities	✓					✓	Uncertainties surrounding the terms of contract
Adverse effect on project performance	✓	✓	✓	✓		✓	Uncertainties of ability to deliver complete project.

9.2.2 Client's financial risk exposure

The advice of professionals is relied on routinely during development projects and share and asset transactions. In the event that the advice is inaccurate, it can have significant implications on the financial risk exposure of all parties. Inaccuracy may stem from the negligence of an adviser

Clients should be aware of the limits within which their advisers operate. Whilst the provision of greater resources would undoubtedly improve the quality of many advisers' work, many uncertainties will always remain. For example, changes in statutes, undiscovered contamination or improved knowledge of the harmful effects of substances. These uncertainties can affect the reliability of an adviser's findings.

Liabilities

The discovery of contamination can lead to escalating project costs and/ or late completion and can have a negative impact on reputation.

Project performance

Clients may have their own contractual obligations to the ultimate owner or occupier of a given development. They may themselves be subject to contractual penalties, liquidated damages and increased legal fees. Delays in project completion can also affect cash flow forecasts and in extreme circumstances jeopardise the viability of an organisation.

9.3 MANAGEMENT OPTIONS

The principal risk management objectives of professional advisers is to ensure that they only take on those liabilities that rightly rest with them given the nature of their involvement in a project – including whether or not they have a stake in its ultimate financial performance. On the other hand, clients seek to minimise their liabilities by transferring them to their advisers through the terms of engagement.

In all cases it is recommended that the detail of terms of engagement are assessed on every occasion and that unacceptable clauses are challenged.

9.3.1 Adviser's options

An adviser's risk management options comprise:

- ensuring the company adopts good working practices (loss prevention)
- reporting the limitations of the work (loss prevention/loss mitigation)
- clearly defining contractual terms, including liability exclusions (control transfer)
- insuring against claims (financial risk transfer).

In addition, financing mechanisms, partnering and profit sharing arrangements may be used to manage the risk.

Good working practices

This strategy seeks to minimise the likelihood of a negligence claim by ensuring that the adviser is working to a high standard. There is no recognised code of practice setting out the standard of care to be applied to specific types of environmental investigation, assessment and remediation works. In addition to the comprehensive CIRIA twelve-volume study, *Remedial treatment of contaminated land*, Special Publications 101–112 (Harris et al, 1995–1998), the Forum on Contamination in Land (FOCIL) maintains a bibliography of current and forthcoming technical guidance documents.

Reporting the limitations of the work

Setting out the limitations of the work in a report enables the client to make more informed decisions regarding the level of reliance that should be placed on the findings. The limitations should be included as caveats, covering technical aspects including uncertainties.

Contractual terms, including liability exclusions

The guiding principle for all professionals is that they must use reasonable care, skill and diligence in the work they have agreed to undertake. Advisers should clearly define the terms of the contract, including obligations, covenants and exclusions. Parties to a contact may exclude liability, although, in reality, commercial pressures often determine which contractual provisions are acceptable.

Clear descriptions of the extent and scope of work can avoid possible ambiguities that may cause disputes between advisers and their clients. A consultant should be clear about what the duties are to which he has been appointed, so he can put in place the services and procedures to deliver services of appropriate quality, and manage client expectations.

Advisers should seek indemnity for losses or expenses incurred which do not arise from their negligence. They should also provide for charging interest for late payment

If possible, advisers should place an upper limit on contractual liability, preferably no more than the available insurance cover or the value of the work being undertaken. To be effective, exclusion clauses that limit financial liability have to satisfy the test of "reasonableness" under the Unfair Contract Terms Act 1977. Set too stringently, the clauses will leave the consultant with no contractual protection. Exclusions should be detailed in the contract rather than the report (which is a post contractual document and bears no contractual effect).

Insurance

Before giving any advice, advisers should check their PI coverage, since insurers often exclude cover for advice on environmental issues. Advisers should ensure that they revise terms and conditions of appointment to exclude risks for which professional indemnity cover is not held. If this is unacceptable, it may be possible to arrange insurance on a "one-off" basis for a particular project, at the expense of either client or the adviser. If cover is not available it may be prudent to decline an instruction.

Client should be aware that an adviser's PI policy will not necessarily provide cover for environmental liabilities. For example, if a client wants assurance that it will not face liabilities from contamination at a site, it should seek alternative methods of protection such as specialist insurance products.

Third parties

Advisers should insert an appropriate clause in the report to limit reliance on the report to the original client and/or other specified parties. An adviser's report is often passed on to third parties who may rely on the findings and lay claim to being owed a duty of care. Where there is express knowledge on the part of the consultant that this is the case, the duty of care owed by the adviser may extend to the other party. Legal advice should be obtained on the effect of any clauses seeking to exclude or limit liability.

An adviser's liability arising under the tort of negligence may not be limited to the parties to the contract. A duty of care may be owed to a third party who suffers economic loss as a result of relying upon an adviser's negligent advice.

In establishing that a duty of care exists, the following criteria must be satisfied.

- it must be foreseeable that damage or loss could result from the negligent advice
- it is just and reasonable to impose liability on the adviser.
- a sufficient "relationship of proximity" must exist between both parties: proximity is an unclear concept, although guidance was provided [*Morgan Crucible Co plc* v *Hill Samuel Bank Ltd, 1990 All ER148*]:
  - the advice must be required for a purpose which is made known to the adviser when the advice is given
  - the adviser knows that the advice will be communicated to a person who may be a specific person or a member of an ascertainable class, and that it will be used by that person for a particular purpose.
- the duty of care is acted upon by that person to his detriment.

9.3.2 Client's options

A client's risk management options comprise:

- clearly defining the contractual terms, including liability exclusions (control transfer)
- insuring against claims (financial risk transfer)
- assessing the financial standing of the consultant (loss mitigation)
- requiring a parent company guarantee (loss mitigation)
- developing a good working relationship with advisers (loss prevention)
- appointing contracts on the basis of value for money rather than necessarily the lowest tendered price (loss prevention).

Clear contractual terms, including liability exclusions

Clients should clearly define the terms of the contract, including obligations, covenants and exclusions. Clear descriptions of the extent and scope of work can avoid possible ambiguities, which may cause disputes. A client should be satisfied that the adviser is clear about the duties to which he has been appointed, and can put in place the services and procedures to deliver services of appropriate quality.

Collateral warranties

Collateral warranties reduce a client's risk exposure by offering the right to sue the adviser for contractual breach i.e. the warranty extends the privity of contract.

Insurance

An adviser's PI will only pay out in a limited set of circumstances. A much wider set of circumstances could result in the client facing additional expenditure due to contamination. Insurance policies are reviewed in Chapter 11.

Good relationship with advisers

Clients should seek to develop good working relationships with their advisers so that despite the contractual nature of the relationship a team spirit is engendered such that the sum is greater than the parts.

Value for money

Clients should resist the temptation to appoint advisers simply on price, using instead criteria based on value for money. Further guidance is given in CLR 12 (DETR 1999a).

Table 9.2 *Examples of performance indicators for technical advisers*

Type of adviser	Performance indicator	Comments
Laboratory	Accreditation of analytical methods eg UKAS	Ensure accreditation is for specific analytical process
	Proficiency testing scheme eg CONTEST	Ask to see their performance charts eg Z scores
Site investigation contractors	Membership of the Association of Geotechnical and Geoenvironmental specialists (AGS)	www.ags.org.uk Client guide to site investigation
Geotechnical specialists	Listing in British Geotechnical Society Directory	The BGS has merged with the ICE and a new system of denoting geotechnical specialists is being proposed The directory is available at www.geotech.co.uk
Drilling contractors	British Drilling Association accreditation	
All	Quality Management System eg ISO 9001	Ask to see named individuals with specific qualifications
	Environmental management systems eg ISO 14001	
	Individual staff qualifications and experience	

SECTION C – RISK FINANCING TECHNIQUES

10 Retained risk financing and composite structures

10.1 INTRODUCTION

Under retained risk financing, the financial consequences of a loss are borne by the organisation experiencing the loss (the risk-holder) or by a group of companies to which the organisation belongs. This differs from financial risk transfer (eg insurance) where another party meets the costs of a loss – see Chapter 11. Full retention and complete transfer are rare – risk financing normally encompasses a blend of retention and transfer.

The decision to retain risk depends on a number of factors, including tax advantages, the expected size, frequency and variability of losses, the size of the business (eg cash flow, profit and net worth), statutory controls and a company's perception of the risk.

10.1.1 Funding mechanisms

Funding arrangements to pay for losses generally fall into four categories:

- direct funding from the "bottom line" where the risk-holder funds losses from its profits without necessarily making any prior arrangements – **see Section 10.2**
- finite risk instruments ie funding arrangements which spread the costs of risks across time – **see Section 10.3**
- captive insurance which spreads the costs across risk exposures – **see Section 10.4**
- composite structures which combine self funding arrangements with risk transfer – **see Section 10.5.**

Retained risk financing can also be divided into two categories according to the source of funds – the first category involves the use of a company's internal resources to pay for losses, the second relies on external finance, in the first instance, which is then repaid. For external funds, the approach may be as simple as arranging a standby credit facility or an overdraft that an organisation can draw on in the event of an unexpected loss. Such committed capital would normally be provided as debt or hybrid debt (redeemable preference shares or mezzanine financing) but may take the form of shares.

Financing techniques may be subject to tax inefficiencies, ie the expenditure may not be tax deductible as a normal trading expense. This will be dependent on a company's specific trading activities and taxation position. In cases of uncertainty, accounting advice should be sought.

10.1.2 Speculative risk retention

Certain contractors and other organisations have been assessing the scope in the market to invest in contaminated sites and obtain the benefits of potential savings or increased value through "working" the asset. For example, an organisation holding land with a contamination risk may invest in the site to secure its remediation as part of a development scheme, which subsequently yields financial benefits. Through this route, a potential liability could be turned into an asset of greater value.

10.2 FUNDING FROM THE BOTTOM LINE

An organisation may choose to finance its losses directly by:

- charging the loss as a direct call against the company's operational costs, ie profit and loss
- selling assets to reimburse the losses incurred by the company
- budgeting for losses and establishing these as provisions against the company's trading activities, thereby providing an available resource (sometimes referred to as a contingency) to fund losses.

In all of these cases, the organisation self-finances its losses or liabilities without necessarily setting up a structure to accumulate funds. A development on this approach is the rolling over of reserved sums from one year to the next. Depending on the accounting arrangements, this may become an accrual fund (see Section 10.3.1).

The purpose of a contingency fund is to set aside a financial "buffer" to pay for unforeseen liabilities. The public sector has historically used contingency funds to manage financial losses. Contingency arrangements may sometimes be pooled between companies, with each member of the pool contributing to a fund which is used to pay for the liabilities of its members. Unlike insurance, if the pool is insufficient to finance liabilities, additional contributions are required from the members of the pool. Contingencies may be applied to prospective or retrospective losses. Case law indicates that a provision may not be treated as a tax-deductible expense to the business unless it can be accurately or reasonably scientifically forecast against a specific liability.

For contamination risks, techniques to protect profit and loss are normally used passively rather than actively. This arises as a result of the risk-holding organisation being unaware of the risks and possible losses to which it is exposed. When a loss occurs, the organisation might discover that its general risk management control measures do not apply to contaminated land issues, so the organisation must finance the loss from the bottom line.

10.3 FINITE RISK INSTRUMENTS

Finite risk instruments are funding arrangements that spread the cost of risk over time. Funds may be held as cash or readily realisable assets which can be used to pay for losses prospectively (ie funds are accumulated in advance of losses) or retrospectively (ie arrangements are made for funds to be made available to pay for losses after they have occurred).

Funds may be held and administered internally by the risk holder, or externally through the appointment of a third party (such as a fund manager). Escrow accounts are administered by independent organisations (usually solicitors), which release funds to pay for losses, or return the fund to the provider only if certain pre-defined conditions are met. Funds might be placed in escrow in the event that a party to a transaction requires funds to cover an identified liability for which the costs are uncertain. Escrow accounts tend to be used for short-term risks or specific issues in relation to contaminated land remediation. They are not generally used for long term risk financing of potential unknown liabilities.

10.3.1 Accrual funds

Accrual funds include provision accounts, reserve funds or sinking funds (sometimes referred to as site pensions). These forms of financing are based on the accumulation of a capital sum to pay for losses prospectively. Funds are structured so that the estimated maximum loss is less than the accumulated reserves.

Accrual funds are held in "earmarked" or "liability" accounts, established specifically to hold reserves against which the potential liabilities might be charged. In the event of a loss, the impact is deducted from the account. An alternative structure involves purchasing an asset that can be readily realised in the event of a loss. This type of funding arrangement is termed an asset account and is suitable for potentially significant losses.

Figure 10.1 illustrates an example of an accrual fund that has been running for a number of years, building up a capital sum of £150 000 which is subject to a claim of £130 000. The fund might continue to accumulate after that time assuming that the claim and works to redress the claim do not alter the need to continue to establish a fund.

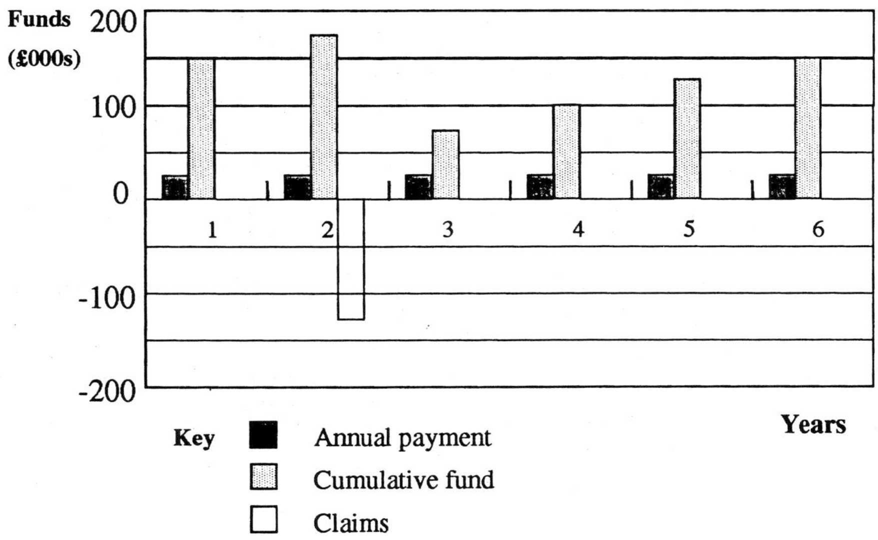

Figure 10.1 *Establishing an accrual (reserve) fund*

Accrual funds are used for contamination risks, particularly in the waste management sector where site pensions are often established to finance the long-term liability risks of landfill sites.

Accrual funds may be subject to tax inefficiencies if the risk is not a reasonable one for which to make a provision. However, where a fund is established for a specific purpose and for a liability that has been pre-determined, it is generally reasonable to deduct funding payments from the normal operating profits of the company.

10.3.2 Bonds

Bonds are financial instruments used to raise an immediate capital sum for a borrower to pay, for example, for a loss or works to prevent future liabilities. Their worth is normally underpinned by the backing of a bank.

Companies providing funds to borrowers are issued with bonds which certify the terms and period by which they will receive a return on their capital plus a returned investment. The terms set by the "bondholder" reflect, amongst other factors, the credit rating of the borrower and the comparable investments available to the bondholder. The structure of a bond is normally set for a limited period of time, typically three to five years. In certain cases, the principal may be amortised.

Bonds are particularly suitable for financing potentially large risks, or for circumstances where there are large uncertainties in the early stages of holding a risk but where the risks decrease with time. This may be the case, for example, where remediation of contaminated land ensures that future losses from contamination are greatly reduced.

10.4 CAPTIVE INSURANCE

Captives are insurance companies owned by one or more insureds or others wishing to insure or reinsure the risks of their companies and/or their affiliates. Many of the principles of general insurance apply to captives.

There are two main types of captive:

- captive reinsurers underwrite risks through a fronting company, which is generally a conventional, external, commercial insurer. The parent or its subsidiaries pay premiums to the fronting company in exchange for cover which is provided on the understanding that a large proportion of the total risk is reinsured with the captive reinsurer
- direct writing captives underwrite risks without using a third party fronting company, generating significant reduction in costs compared to captive reinsurers.

Captives have been used increasingly as a general vehicle for insurance funding and in some circumstances for specific environmental risk coverage (Lockett, 1996). Those companies that have set up captive insurers cite a number of reasons for doing so, including (Datamonitor, 1995):

- captives can cover risks which might otherwise be uninsurable through the conventional insurance market
- there is opportunity to generate cost savings and increased profitability. The main opportunity stems from the parent company no longer having to pay a premium that includes an underwriter's expenses and an agent's fees. A further source of profitability is the investment income which is retained by the captive as opposed to being earned by the insurer. Often, this income is enhanced by the tax benefits of locating the captive in an offshore domicile
- policies are flexible, for example, coverage can be tailored to a company's specific needs whilst payment terms can be arranged to suit the client (which in turn brings about benefits in cash flow management)
- the control of risk is enhanced. An advantage of captives lies in the capacity to centralise control of an organisation's insurance strategy, whilst at the same time allowing autonomous business units to take out customised policies. Some companies have also found that captives help to control and co-ordinate foreign and domestic insurance programme
- the level of confidentiality provided by a captive may be higher than that of a conventional insurer

Examples of two captives are described in Box 10.1.

Box 10.1 *Examples of captive arrangements (adapted from Datamonitor 1995)*

Hoechst/Bayer/BASF

Early in 1994 the three major German chemical manufacturers Hoechst AG, Bayer AG and BASF AG established Indurisk Re – a group reinsurance captive initially providing the three companies with insurance cover for environmental liability. The captive was set up for a number of reasons:

- tightening German and EC legislation which had increased the degree to which companies could be held liable for environmental damage
- a general absence of environmental liability insurance in Germany at that time
- significant increases and fluctuations in insurance premiums for large commercial risks
- a DM 425 million loss suffered by BASF, which had increased premiums to a level where captive insurance was a more economical option.

Indurisk Re was located in an offshore domicile (Luxembourg) which, being independent of German fiscal jurisdiction, brought tax benefits (subject to local regulations at state level) and more stable premiums. Each member of the captive self-insured losses of up to DM 20 million, with the remaining 80 per cent of the risk being insured with the captive. No third party direct insurer was used to front the policy.

BP

While most companies possessing a captive insurance company tend to self-insure small, frequent risks rather than larger, more predictable ones, BP adopted the opposite strategy. The company chose to insure by conventional insurance those risks causing losses of less than $10 million, while self insuring those with a potential for generating claims of between $10 million and $500 million. This strategy was driven by a number of factors, including:

- an audit of insurance expenditure in the 1980s which showed the company had spent $1.2 billion on insurance premiums and only made claims of $250 million
- the fact that BP had a better credit rating than most of the underwriters with which it had been dealing at the time
- a recognition that BP was in a better position to assess and price its specialist risks than insurers.

BP opted to insure smaller risks via conventional insurance because the competition amongst insurers was greater at this level and insurers were accustomed to handling large numbers of small claims. The company chose to fund large losses by raising money from the capital markets – a move that was facilitated by the size and financial standing of BP.

10.5 COMPOSITE STRUCTURES

Composite structures combine self-insured retention arrangements with risk transfer. These are particularly useful in managing the risks of known liabilities for which the costs and timing are uncertain. Such risks may not be insurable. A model example of a composite (ART) structure involving a finite risk instrument and insurance is illustrated in Figure 10.2 and described in the text beneath. This example serves to illustrate the potential benefits of composite structures, ie:

- earnings can be protected and cash flow stabilised
- losses from the balance sheet are "smoothed" over a period of years
- premiums can be deducted against tax
- if the loss does not occur, the client receives a substantial proportion of the profit made on the fund
- at the end of the contract, if there is a positive fund it can be used to extend the programme, offsetting future premium costs or increasing the scope of the programme.

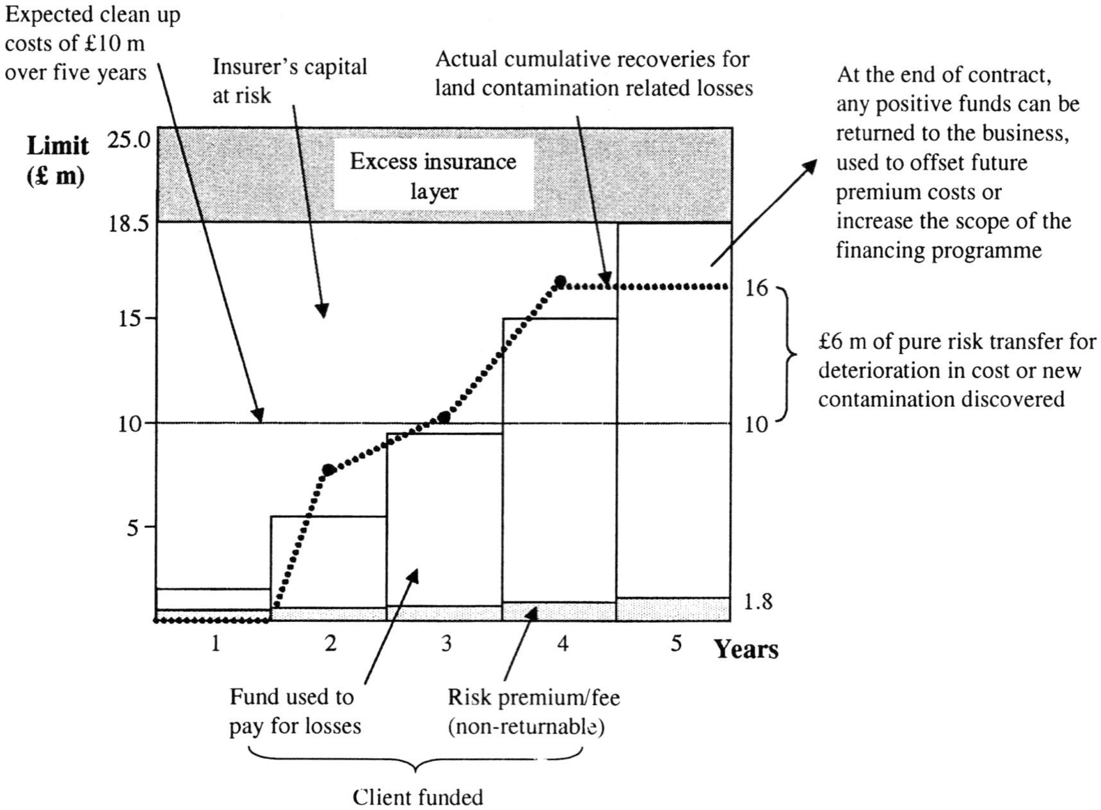

Figure 10.2 *Model of a composite financing arrangement*

In this example, an organisation estimated that its contingent liabilities were £10 million over a five-year period. The company entered into a five-year contract to capitalise an "experience account" managed by another organisation (eg an insurer). This account provided the principal source of capital for financing losses. For simplicity it was assumed that the losses would be paid out within one calendar year. In reality, money would probably flow out from the insurer on a slower pattern which would enhance the value of the experience account.

The total insurance limit issued by the insurer was £25 million against expected costs of £10 million and a premium paid of £18.5 million. This provided acceptable risk transfer for tax and accounting regulations. The entire fund limit could, in theory, have been paid out on day one of the contract despite the fact that the premium payments were spread over five years. Thus, the "timing risk" of the liability was borne by the insurer, not the client who had total certainty of timing and cost for the duration of the contract.

Ninety percent of the premium paid was ceded to the experience account. This was held on behalf of the client for use solely to pay for losses. An investment income of 6 per cent applied to the experience account over the funding period, and this accrued to the benefit of the client. By doing this, the programme was able to take into account the time value of money.

Ten percent of the premium paid was paid to the insurer to cover the catastrophe cover excess of the "experience account" fund. In this example the pure risk transfer granted was £6 million but in addition the insurer took the timing risk of payments not accruing as expected over the contract. These percentages are for illustration only and the actual values will depend on the limit required, the certainty of payment pattern and whether coverage is also granted for prospective contamination as well as the unexpected additional cost of known contamination.

In return for its financial outlay, the company achieved:

- stability of costs and elimination of timing risk
- transfer of balance sheet volatility to the insurer
- £6 million of pure risk transfer for deterioration in cost or new contamination discovered
- tax deductibility of premiums paid into the experience account
- a strategic response to corporate governance requirements and FRS12 accounting regulations
- a return of up to 90 per cent of premium would have been paid in the event that no losses occur during the contract period.

11 Insurance

11.1 INTRODUCTION

UK industry has relied, and in many cases continues to rely, on traditional liability insurance, such as public liability (PL) and property policies, to provide indemnity for contamination-related losses. However, not all of these financial risk exposures are effectively covered by traditional policies. In some cases the position is clear – policies will either trigger or not in the event of a specified loss (so long as various conditions in the policy are satisfied). For other risk exposures, the position is less clear and this is unlikely to change until the meaning of key policy terms have been tested in the UK courts. In the meantime, organisations relying solely on traditional liability policies may be exposed to uninsured or under-insured contamination-related liabilities.

The deficiencies in, and uncertainties of, cover provided by traditional liability policies have been addressed by the introduction of a range of specialist environmental insurance products. These products complement traditional liability policies by replacing or filling in the gaps in these policies, and by providing niche cover for environmental liabilities (Fogleman, 1999a). Specialist insurance cover is available for a comprehensive range of financial risks associated with existing (unknown or residual) and new contamination, as described in Section 11.2.5.

This chapter provides a review of the conventional and specialist environmental insurance markets in relation to the financial risks of land contamination in the UK. The specialist environmental insurance market has undergone significant change since the early 1990s, both in terms of the nature of policies and the number of companies offering specialist products and solutions. The experience of insurers in the US may provide the best guide to future trends in the UK, although there are marked differences in the legal regimes operating in these countries, the structure of the court systems and the courts' approaches to insurance coverage matters.

11.1.1 The US experience

Comprehensive/commercial general liability (CGL) policies, which are somewhat similar to PL policies in the UK, had until the mid-1970s been written mostly without pollution exclusion clauses. Concerned by the problems of asbestos and increasing environmental liabilities, insurers initially placed restrictions on cover for gradual pollution. The exclusion clauses related to the cause of the damage rather than the damage itself. However, it soon became clear that the cover was open to several interpretations. As a result, an absolute pollution exclusion clause was inserted in CGL policies, although even that clause has been subject to legal challenge and debate (Fogleman, 1999b). In turn, this encouraged the development of pollution-specific insurance products.

11.1.2 The insurance market for environmental risks in the UK

The insurance market for UK risks reacted to the US experience and increasing environmental liabilities in the UK by limiting cover in virtually all PL policies written since 1990 to sudden and unintended pollution incidents. The basis of cover results from the use of certain exclusion wordings, the most common of which are the model exclusion issued by the Association of British Insurers (ABI) and clause 1685 issued by Lloyd's Non-Marine Association (see Box 11.1). To date, no pollution exclusion clause has been tested in a court in the UK.

Box 11.1 *Examples of pollution exclusion clauses*

ABI Model Pollution Exclusion Wording

"This policy excludes all liability in respect of Pollution or Contamination other than caused by a sudden identifiable unintended and unexpected incident which takes place in its entirety as a specific time and place during the period of insurance.

All Pollution or Contamination which arises out of one incident shall be deemed to have occurred at the time such incident takes place.

The liability of the Company [Insurer] for all compensation payable in respect of all Pollution or Contamination which is deemed to have occurred during the Period of Insurance shall not exceed £[amount] in the aggregate.

For the purpose of this Endorsement "Pollution or Contamination" shall be deemed to mean

a) all pollution or contamination of buildings or other structures or of water or land or the atmosphere; and

b) all loss of damage or injury or indirectly caused by such pollution or contamination"

Lloyd's Non-Marine Association (NMA) Clause 1685

"This insurance does not cover any liability for:

1) Personal injury or Bodily Injury or loss of, damage to, or loss of use of property directly or indirectly or indirectly caused by seepage, pollution or contamination, provided always that this paragraph shall not apply to liability for Personal Injury or Bodily Injury or loss of or physical damage to or destruction of tangible property, or loss of use of such property damaged or destroyed, where such seepage, pollution or contamination is caused by a sudden, unintended and unexpected happening during the period of this Insurance

2) The cost of removing, nullifying or cleaning-up seeping, polluting or contaminating substances unless the seepage, pollution or contamination is caused by a sudden, unintended and unexpected happening during the period of this Insurance

3) Fines, penalties, punitive or exemplary damages.

This Clause shall not extend this Insurance to cover any liability which would not have been covered under this Insurance had this clause not been attached.'

The specialist environmental insurance market developed in the UK from the early 1990s. Since then, insurers have responded to UK market requirements by offering policies with broader coverage (for example, property portfolios and liabilities arising from future changes of law), less restrictive conditions, higher limits of indemnity (tens of millions of pounds) and longer periods of cover or notification (up to 25 years cover in some cases).

The specialist market is increasingly flexible in its products and solutions. Another feature has been the introduction of modular policies, which allow the insured to choose appropriate cover from a suite of options.

11.2 POLICIES

For all insurance policies, it is essential that the terms and conditions are carefully reviewed in order to be clear on what is, and what is not, covered.

Insurance cover for land contamination risks in the UK may be provided under the following main categories of insurance:

- public liability (PL)
- property
- professional indemnity (PI)
- contractors all risk (CAR)
- specialist environmental.

The main features of these insurance products are described in Sections 11.2.1–11.2.5. An explanation of selected key insurance terms is provided in Box 11.2.

Box 11.2 *Explanation of selected insurance aspects and terms*

Policy basis	Insurance is written on an occurrence or claims made basis. Occurrence-based policies provide cover for losses from property damage or bodily injury that occur during the policy period. Due to the potential for progressive property damage or latent bodily injuries, occurrence worded policies carry a long tail risk of liability for insurers. Claims made policies provide cover only if the insurer is notified of a loss and/or a claim is lodged during the insured period. If no notification or claim is made during this time, the policy will lapse and, if renewed, a new policy will commence. Claims made policies provide the insurer with greater control over the risk. The insured loses the benefit of cover as soon as the policy expires, including cover for events which have occurred during the period of the policy. To mitigate this, claims made policies often include an extended reporting period (ERP). Policies may also allow notification of circumstances before a claim is made. In such cases there is "sunset clause" which places a time limit on when a claim can be made following the notification. This avoids the policy operating like an occurrence form.
Limits of indemnity	The limit of indemnity is the total amount of indemnity payable by the insurer for all claims and losses after application of any deductible. Sub-limits may be set for each and every claim and/or in the aggregate.
Deductible	The deductible is the sum which the insured will retain of each and every claim or in the aggregate of losses and costs incurred in the policy period.
Disclosure of information	Insurance contracts are based on the principle of *uberimmae fides* (utmost good faith) whereby all parties to the contract are legally obliged to disclose all information material to the risk. To ensure that a policy is not voided, it is essential that the proposal form, which usually forms the basis of the contract, is completed accurately, ensuring that it discloses all material facts and contains no statements that would amount to misrepresentation of material facts.

Other insurance may, under limited circumstances, also provide cover for contamination losses. For example, Directors and Officers (D&O) policies may provide cover for defence costs but not property damage or bodily injury. EL policies cover injuries to employees and do not contain a pollution exclusion clause.

11.2.1 Public liability (PL)

Public liability policies provide an indemnity for loss or damage to third party property or bodily injury arising from the activities of the insured. Product liability cover may be contained in PL policies, providing indemnity for losses that may occur as a result of products supplied by the insured. Policies are normally written on an occurrence basis, meaning that the policy provides cover for property damage or bodily injury that occurs during the policy period.

Cover may be narrow in respect of contamination for the following reasons:

- virtually all policies contain an exclusion in respect of sites that are owned or occupied by the insured
- exclusions for gradual pollution are included in most policies written since 1990
- depending on the wording of individual policies, it is uncertain whether the words "damages" or "compensation" in the coverage clause cover remediation costs.

Despite these limitations, PL policies may still provide protection for the insured beyond third party liabilities arising from sudden pollution incidents. PL policies written before the introduction of pollution exclusion clauses may also provide an element of cover if damage occurred during the insurance period and the insured is legally liable.

11.2.2 Property

Property insurance may provide cover for contamination-related losses where these losses are associated with a named peril such as fire. For example, a degree of own-site remediation may be provided coincidentally as part of the works following a fire. This is normally limited to debris removal.

The question as to whether property insurance covers contamination also depends on the list of scheduled property in the policy. This list may not cover land or, in some policies, land may be excluded. Many property policies now exclude cover for gradual or both gradual and sudden contamination.

11.2.3 Professional indemnity insurance (PI)

Professional indemnity insurance covers professionals for claims made against them for negligence. In the case of contamination-related claims, the sums involved can run into millions of pounds (see Box 1.1), far exceeding the capital base of most environmental consultancies. Therefore, without properly worded PI, such organisations could be in a particularly vulnerable financial position.

Pollution exclusion clauses are sometimes present in PI policies, reducing considerably the protection afforded when hiring professionals. For clients relying on the PI cover of their advisers to provide cover for contamination-related losses, there are a number of limitations that they should be aware of in addition to pollution exclusion clauses.

These limitations, which apply irrespective of whether or not the requirement for PI cover is made a condition of the contract with an adviser, are listed below.

1. Policies are normally written on a claims-made basis (see Box 11.2) which means that for the policy to respond, the insured must make a claim during the policy period or any extended reporting period. Clients have reacted to this by requiring their advisers to maintain a certain level of PI for a specified number of years after completion of a project. However, it cannot be guaranteed that the necessary insurance will be available in the future. Depending on claims experience, policy renewal may also be expensive, due to the disclosure obligations which apply when completing a proposal form for insurance (Stubbs, 1998).

2. PI is normally provided on an "each and every" basis, but with aggregate limits on pollution claims (see Box 11.2). Therefore, the insured may exhaust its cover in any one insurance period, leaving the insured with little or no cover during the remainder of this period. Clients should, at the time of procuring the services of an organisation, seek to confirm the level of PI cover that is available, as well as the status of any claims that are pending against the current policy. Clients should also consider requiring contractors on long term contracts to disclose any changes to their PI cover, as and when any such changes occur.

3. Depending on the insurer and policy, the insurer may regard the policy limit as including costs. Therefore, even if negligence is proven (which may be difficult) the legal costs of defending the action may have significantly reduced the funds to pay the claim.

Certain specialist environmental insurers offer PI policies without exclusions.

11.2.4 Contractors all risks

Contractors all risks (CAR) insurance provides an indemnity to the insured for loss, damage or destruction by any cause whatsoever of the contract work. Policies do not cover third party liability and are normally subject to the same pollution restrictions as PL policies.

11.2.5 Specialist environmental insurance

Specialist environmental insurance – sometimes loosely referred to as environmental impairment liability (EIL) insurance – is offered by a number of companies in the UK, including:

- AIG Europe (UK) Limited
- Certa underwriting on behalf of Allianz Cornhill International
- Chubb Insurance Company
- ECS underwriting on behalf of XL Europe Limited
- Independent Insurance Company
- NHBC
- Reliance National Insurance Company (Europe) Limited
- R J Wallace & Others Syndicate (Lloyds Syndicate)
- Royal & SunAlliance
- Swiss Re
- Zurich Building Guarantees
- Zurich Specialities London.

Some of these insurers offer a broad range of products, whilst others specialise in a more limited range of policies. Other insurers may also provide some form of environmental insurance on a policy-by-policy basis.

The following types of specialist environmental insurance policy are available in the UK. Depending on the insurer, the policy and the risk insured, cover may be extended to include business interruptions, such as increased operating costs, loss of profit and reduction in revenues.

In certain cases, specialist policies from the same insurer may be combined to provide more comprehensive protection. This also avoids issues arising as to whether a loss is covered by one policy and not another.

Third party liability

These policies provide the insured with cover for third party liabilities in respect of bodily injuries and property damage arising from sudden and/or gradual pollution. Depending on the policy, remediation costs, defence costs, certain specified economic losses and the costs of taking measures to avoid an imminent loss may also be covered.

Own-site remediation

These policies indemnify the insured for the costs of statutory enforced remediation of contamination on their own site.

Combined third party liability and own-site remediation

These policies combine the cover provided by individual third party and own-site remediation policies into a single policy.

The above policies are written on a claims-made basis and, in most cases, on a site-specific basis. Cover is generally limited to a maximum of five years (one to three years is more typical) with up to three years retroactive cover.

Property transfer risks

These policies provide cover for "pre-existing" unknown conditions on a site. Known contamination may also be covered where information is provided by the insured and the risks associated with the known conditions are acceptable to the insurer.

Policies are written on a claims made basis and a retroactive date is normally not specified. Depending on the insurer and risks being insured, policies are available for periods of up to 25 years for a one-off premium. The insurer may retain the right to re-evaluate the risk at pre-set times during the policy in order to trigger a claim. Depending on the insurer, cover can be provided for:

- single sites or land portfolios (split aggregate limits are normally offered for portfolios)
- contamination that renders the site unsuitable for use
- the costs of taking voluntary action to avoid legal liability
- economic losses arising from business interruption, eg loss of rental stream or operating revenue
- the costs of defending a criminal charge brought on by an insured's activities that result in pollution.

It is important to consider the relevant policy clauses, particularly those relating to:

- cancellation
- periods of changes of legislation – up to 10 years may be offered, but this may be extended for additional premium
- discovery of contamination by the insured
- changes of use of the site.

Common exclusions on property transfer, third party liability, own-site remediation and combined policies include: radioactive matter, asbestos, lead-based paints, underground storage tanks (USTs), endocrine disrupters and known conditions. Exclusions are negotiable and coverage may be extended to include these risks. This is usually addressed on a policy-specific basis, subject to underwriting and certain other conditions being satisfied, for example, agreeing to a programme of leak detection for USTs.

Each of the above types of policy has a deductible and an aggregate limit.

Contractors pollution liability (CPL)

CPL insurance provides coverage for sudden and gradual pollution liabilities arising from the activities of contractors (remediation, construction and engineering contractors etc). Indemnity may extend to the cost of remediating both the site on which the contractor is working and neighbouring land, the cost of works required by the enforcing authority and defence costs. Completed operations coverage is also available.

Policies are written on a claims made basis, or less frequently on an occurrence worded basis. For claims made policies, there is usually an option for an extended reporting period. In certain cases, the insured's clients can be covered as additional insureds under the policy. Alternatively, a client may purchase cover on behalf of its contractor. Contractors may purchase blanket cover or cover for a specified project. Operations are specified on the policy.

CPL policies are also provided for companies, such as environmental consultants and laboratories, which may engage in contracting activities. Policies are also available that combine the insuring agreements of CPL and PI. This combination provides cover for sudden and/or gradual pollution-related acts, errors or omissions resulting from professional and contracting services specified in the policy.

Clean-up cost cap/remediation stop loss

Clean-up cost cap policies indemnify the insured for remediation costs, as defined in the remedial strategy, that exceed a cost estimate agreed with the insurer. This value may be defined by modelling the remedial plan in order to express technical risks in financial terms. The policy attaches over a prescribed self-insured retention (SIR), which is generally equal to the expected costs of remediation plus a buffer layer (Figure 11.1.)

Policies respond to increased projects costs caused, for example, by the discovery of unknown contamination or greater amounts of known contamination. Coverage is also provided for extra costs to change the remediation works, if the enforcing authority requires these changes. Cover may also be available for additional site investigation and any legal costs that could not have been reasonably anticipated under the remediation plan.

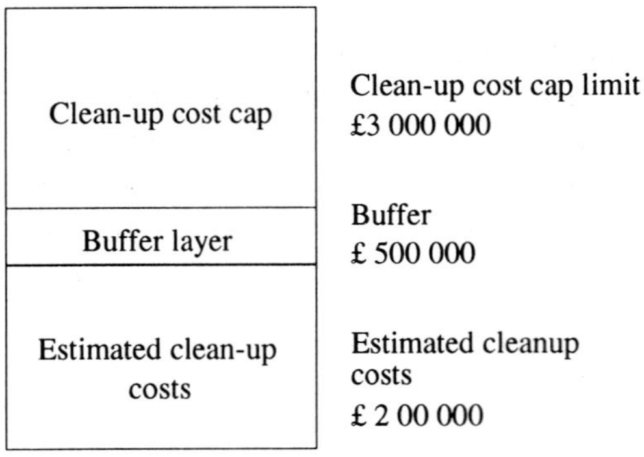

Figure 11.1 *Example of clean-up cost cap*

The buffer that is set will reflect the insurer's confidence in the insured's initial cost estimates and contingencies. The premium will normally depend on the cost and type of remedial work, the comprehensiveness of the remedial strategy and the amount of self-insurance. Insurance may be included within a profit and risk sharing arrangement, as illustrated in the example in Box 11.3.

Box 11.3 *Example of a risk-sharing contract with a cost cap insurance element*

Risk-sharing contracts can be structured to incorporate an insurance element, as shown in Figure 11.2.

In this example, the remediation costs of a development contract are expected to be £10 m. The contract incorporates an excess layer of insurance which covers cost overruns of up to £20 m for a premium of £1 m. The £1.5 million insurance deductible is covered in full by the contractor, which puts all of its profits at risk. The site owner bears the unlikely risk that costs exceed the insurance cover (ie any project costs in excess of £30 m), whilst sharing any cost savings (if the project costs less than £10 m) with the contractor according to a predetermined formula.

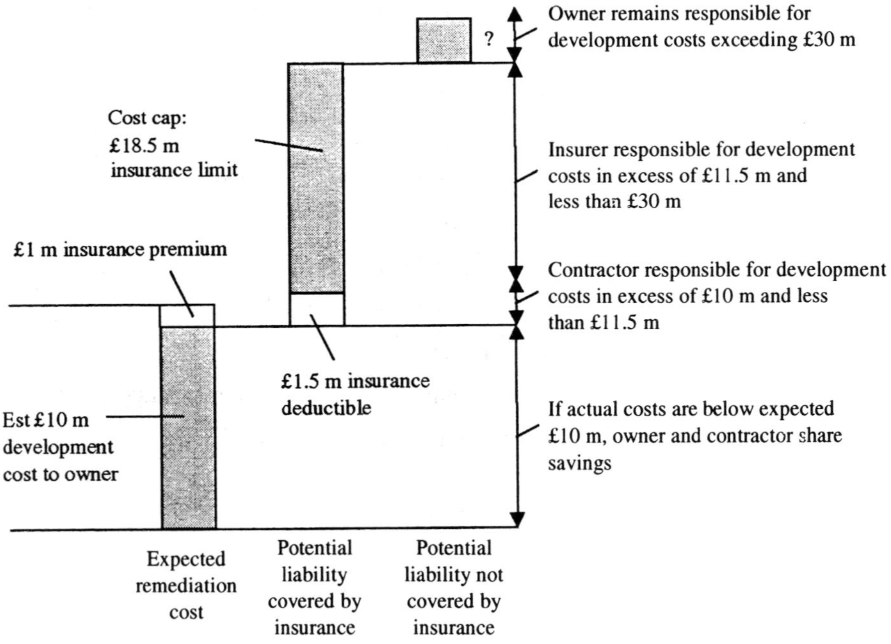

Figure 11.2 *Example of a risk shared contract incorporating insurance (modified from Merkl and Robinson, 1997)*

Legal indemnity

Insurance policies are available on a claims made basis to cover payments made and the legal costs incurred in defending claims under a warranty or indemnity. Some insurers will "back to back" their policy wording with that of the warranty or indemnity subject to a cap in time and amount if no such limits are specified in the warranty/indemnity. Alternatively, specific exclusions may be written into the insurance policy for any uninsurable clauses in the warranty/indemnity. Depending on the insurer, policies are available for periods of up to 25 years.

Lender liability policies.

Lender liability insurance is designed to protect a lender's loan portfolio (single sites may also be covered) against losses incurred as a result of contamination. Policies often contain a dual trigger to insure the loan and indemnify the insured for third party liabilities which may be incurred, for example, as a mortgagee in possession.

In the event of a loss arising from a default by a borrower on a loan facility secured by an insured property, the insurer will usually pay the lesser of the outstanding balance or the latest declared insured value (as agreed by the insurer). Depending on the policy, the profit on a loan may also be covered.

Cover for third party liability may extend to bodily injury, property damage and clean up resulting from on-site and off-site pollution. Legal expenses and defence costs may also be covered in the policy.

Policies are written on a claims made basis with an automatic and/or optional extended reporting period, aggregate limits and deductibles.

11.3 POLICY SELECTION CRITERIA

The main policy characteristics that should be considered when selecting appropriate insurance include:

- form (claims made or occurrence wording)
- period (duration of policy, straight cover or extended reporting periods)
- coverage, for example:
  - the events that are covered
  - the perils that are covered
  - the losses that will be reimbursed
  - the locations that are covered
  - the persons who are insured
  - the nature of any exclusions
  - the nature of any conditions
- limits of indemnity (each and every claim, aggregate or no limit)
- terms (price of insurance and deductibles).

In addition, it is important to consider the following three factors.

Security of insurer

Depending on the financial standing of the insurer, there remains a credit risk that a claim will not be paid. Security ratings are provided by a number of organisations, such as Moody's Investors Services, A.M Best and Standard and Poor's Corporation. These rating systems are based on an assessment of the capital backing and nature of a business. Generally, those organisations that possess secure, well diversified and well understood business portfolios are less susceptible to risk and are given higher ratings. This in turn reflects the level of confidence that an insurer will meet its payment obligations.

Business benefits

Insurance does not just indemnify against losses suffered by the insured. It also reduces the gap between the actual and perceived cost of the risk by reducing the level of financial uncertainty. In turn this increased certainty can provide potential business benefits by facilitating transactions, avoiding potential "deal breakers" and helping to realise the maximum value of an asset.

Information requirements

Site surveys are not always necessary to obtain cover for certain financial risks. Reliance is often made on available site reports (audits, site investigations etc), telephone surveys, site inspections by the insurers or commercially available information databases.

References

Accounting Standards Board (ASB) (1998)
Financial reporting standard 12: Provisions contingent liabilities and contingent assets
ASB Publications, Milton Keynes

Association of British Insurers (1998)
Joint pollution working group – recommendations for the underwriting of pollution risks
Association of British Insurers, London

Barber, J N (1993)
Collateral warranties
Association of Geotechnical Specialists,

Braithwaite, P (1999)
Physical remediation methods
Remediation technologies workshop, Royal Marsden, 6 July 1999
IBC, London

Catalano, A and Frampton, R (2000)
Finance Special: Securitisation
Estates Gazette, 13 May 2000

CLAIRE (1999)
Contaminated land – the legal framework
CLAIRE View, Winter 1999, pp 1–8

Confederation of British Industry (1993)
Firm foundations
CBI, London

Datamonitor (1995)
Captive insurance
Pearson Professional Ltd, London

Department of the Environment, Transport and the Regions (1994a)
Documentary research on industrial sites
Contaminated Land Research Report *(CLR) 3*
Her Majesty's Stationery Office, London

Department of the Environment, Transport and the Regions (1994b)
Framework for contaminated land: outcome of the Government's policy review and conclusions from the consultation paper "Paying for our past"
Her Majesty's Stationery Office, London

Department of the Environment, Transport and the Regions (1994c)
Planning policy guidance 23: planning and pollution control
Her Majesty's Stationery Office, London

Department of the Environment, Transport and the Regions (1999a)
A quality approach for contaminated land consultancy
Contaminated Land Research Report (CLR) 12
Her Majesty's Stationery Office, London

Department of the Environment, Transport and the Regions (1999b)
Model procedures for the management of contaminated land: risk assessment.
(Monitor Environmental Consultants and Stanger Science and Environment, in prep.)

Department of the Environment, Transport and the Regions (2000)
Circular 2/2000 Contaminated land: implementation of Part IIA of the Environmental Protection Act, 1990
Her Majesty's Stationery Office, London

Derbyshire County Council (DCC) (1986)
Report of the non-statutory public inquiry into the gas explosion at Loscoe, Derbyshire Volumes I and II
DCC, Derby

Doolittle, I (1999)
Accounting for environmental costs and liabilities: a lawyer's perspective
Environmental Accounting and Auditing Reporter, June 1999, pp 4–6

DTZ (unpublished)
Money into property 2000
DTZ Publications

ENDS Report (2000)
No 305, June 2000, pp 4–5

Environment Agency (1998)
Enforcement Policy
www.environment-agency.gov.uk

Fogleman, V (1999a)
Specialist environmental insurance
Commercial Environmental Law and Liability (Simon Payne, ed) B8/32–B8/56
Sweet and Maxwell, London

Fogleman, V (1999b)
When petrol is not a pollutant
Insurance Day, 31 August 1999, p 9
LLP Ltd, London

Gardner, D (1999)
Accounting for Acts
Business Risk, March, p 18

Graham, T (1995)
Contaminated land
Jordan Publishing Ltd, Bristol

Harris, M R, Herbert, S M and Smith, M A (1995–1998)
Remedial treatment for contaminated land , Volumes I–XII
CIRIA Special Publications 101–112
CIRIA, London

Hunt and Hilton (1997)
Property companies: an industry accounting and auditing guide, 5th edition
The Institute of Chartered Accountants in England and Wales

Institution of Civil Engineers and the Faculty and Institute of Actuaries (1998)
Risk analysis and management for projects
Thomas Telford Ltd, London

John, A (1999)
Lessons from the marketplace – common pitfalls in business transactions
Contaminated land: preparing for Part IIA, 9th Annual Conference,
25–26 January 1999
IBC UK Conferences Limited, London

Kass, G A (1998)
A brown and pleasant land – household growth and brownfield sites
Parliamentary Office of Science and Technology, Report 117
POST, London

Laidler, D, Bryce, A and Boswall, J (unpublished)
Guidance on the sale and transfer of land which may be affected by contamination
CIRIA, London, Special Publication 99

Lockett, N (1996)
Environmental liability insurance
Cameron May, London,

London Stock Exchange (1998)
Committee on Corporate Governance: *The combined code, principles of good governance and code of best practice*
Gee Publishing, London

Loss Prevention Council (1997)
Assessment of pollution risks
Report SHE 14
Loss Prevention Council, London,

Mallett, H (1999)
Contaminated land – getting it right
Proceedings of the 10th IBC Conference on Contaminated Land
Royal Marsden NHS Trust, London, 5 July 1999

Merkl, A and Robinson, H (1997)
Environmental risk management: take it back from the lawyers and engineers
The McKinsey Quarterly, 1997, No 3

Nathanail, C P (2000)
Remediation of contaminated land,
Environment Business Magazine, May 2000

NATO/CCMS (1985)
Contaminated land: reclamation and treatment
NATO CCMS, Volume 8
Plenum Press (New York and London)

Nicholls, M (2000)
Pension sector shows green tinge
Environmental Finance, May 2000, pp 16–18

Office for National Statistics (ONS) (1999)
Economic trends

Parliamentary Office for Science and Technology (1993)
Contaminated land
POST, London

Petts, J (1994)
Contaminated sites: blight, public concerns and communication
Land Contamination and Reclamation, Volume II, Number 4

Phillips, J and Ferera, A (1998)
Warranties and indemnities in sale and purchase agreements
Corporate Finance Guideline, Issue 9
Institute of Chartered Accountants in England and Wales

Smith, M H (1999)
The New Environment Act: protecting your valuable assets
Conference Proceedings, 26 May 1999

Stubbs, A (1998)
Environmental law for the construction industry
Thomas Telford, London

Sury, M and Slingsby, A (1992)
Proceedings of the 2nd International Conference on Construction on Polluted and Marginal Land
Brunel University

Toulmin, C (1998)
Damage to property – *Blue Circle Cement* v *Ministry of Defence*
Estates Gazette Case Summaries (EGCS), 93, p 1

Turnbull (1999)
Internal control: guidance for directors on "The combined code"
Institute of Chartered Accountants in England and Wales, London

Urban Task Force (1999)
Towards an urban renaissance
Her Majesty's Stationery Office, London

Waite, A and Jewell, T (1997)
Environmental law in property transactions
Butterworths, London

Wickens, D, Runfitt, A and Willis, R (1993)
Survey of derelict land in England, 1993, Volumes I and II
Department of the Environment
Her Majesty's Stationery Office, London

WS Atkins Consultant Ltd. (1998)
CDM Regulations – practical guidance for clients and clients' agents
CIRIA Report 172
CIRIA, London